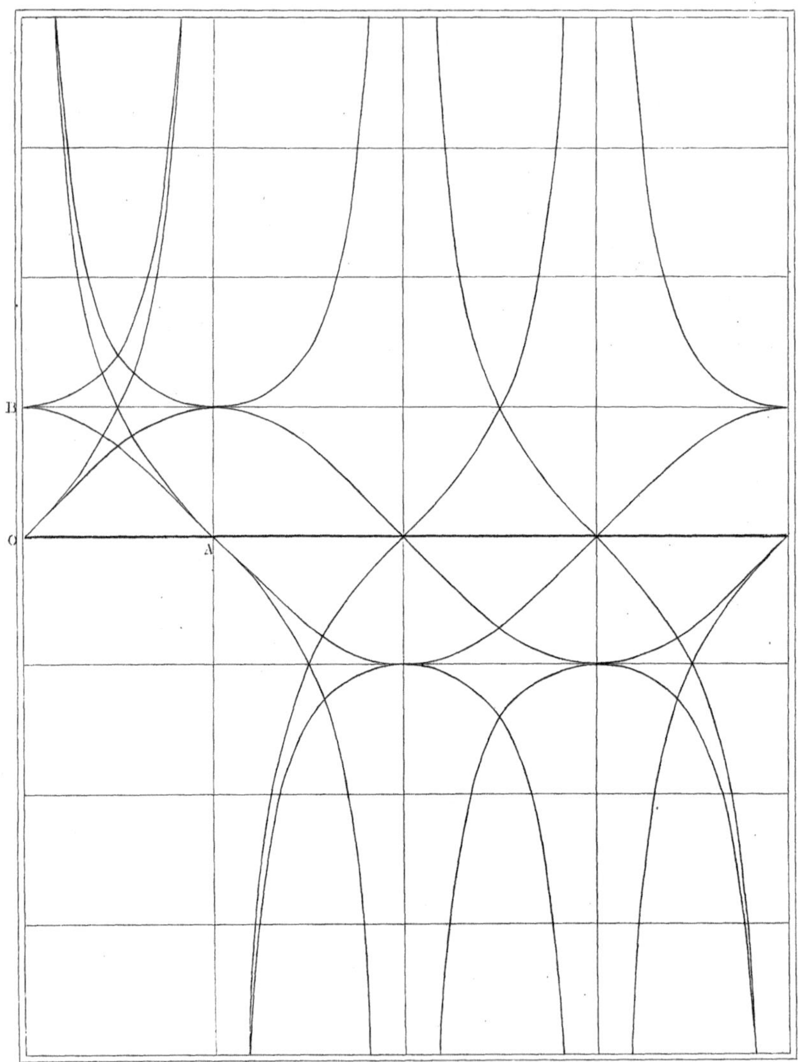

London, Taylor, Walton, & Maberly.

TRIGONOMETRY

AND DOUBLE ALGEBRA

BY

AUGUSTUS DE MORGAN

OF TRINITY COLLEGE, CAMBRIDGE

SECRETARY OF THE ROYAL ASTRONOMICAL SOCIETY
FELLOW OF THE CAMBRIDGE PHILOSOPHICAL SOCIETY
AND PROFESSOR OF MATHEMATICS IN UNIVERSITY COLLEGE LONDON.

La seule manière de bien traiter les élémens d'une science exacte et rigoureuse, c'est d'y mettre toute la rigueur et l'exactitude possible.—D'ALEMBERT.

Tant que l'algèbre et la géométrie ont été séparées, leur progrès ont été lents et leurs usages bornés; mais lorsque ces deux sciences se sont réunies, elles se sont prêtées des forces mutuelles, et ont marché ensemble d'un pas rapide vers la perfection.—LAGRANGE.

LONDON:

PRINTED FOR TAYLOR, WALTON, AND MABERLY,

BOOKSELLERS AND PUBLISHERS TO UNIVERSITY COLLEGE,

UPPER GOWER STREET AND IVY LANE, PATERNOSTER ROW.

1849.

CAMBRIDGE:
PRINTED BY METCALFE AND PALMER, TRINITY-STREET.

PREFACE.

THE work before the reader is entirely new, not being in any sense a second edition of that which I published on the same subject in 1837.

It consists of two books. In the first, I have endeavoured to give the student who has a competent knowledge of arithmetic and algebra—as much for instance as is contained in my works on those subjects, to which reference is made in various places—a view of trigonometry, as a branch of algebra and a constituent part of the foundation of the higher mathematics. In the second, I have given an elementary view of algebra in its purely symbolic character, with the application of that geometrical basis of significance which affords explanation of every symbol.

The term *double algebra* has not yet obtained currency, though that of *triple algebra* has, of late years, been much employed. It means algebra in which each symbol stands for an object of thought having two distinct and independent qualities: just as the symbol of a straight line, to be perfect, must designate both the length and direction of the line. I have not, after much thought, and some discussion, been able to fix on a better name of sufficient brevity. If, by the application of a somewhat startling adjective to the word *algebra,* any of those who are still bewildered by an art in which *impossible quantities,* or quantities which are not quantities, are made objects of reasoning, should become aware that by slow degrees, and the union of many heads, the art has become a science, and the impossibilities possible, they, at least, will have no objection to the phrase.

A. DE MORGAN.

University College, London,
Feb. 10, 1849.

LIST

OF SOME WRITINGS ON THE SUBJECT OF ALGEBRA,

In which the peculiar Symbols of Algebra are discussed.

London, 1685, folio. JOHN WALLIS. *A Treatise of Algebra, both historical and practical.* Reprinted in Latin, with additions, in the second volume of *Wallis's Works*, Lond. 1693, folio.

Naples, 1687, folio. GILES FRANCIS DE GOTTIGNIES. *Logistica Universalis.*

London, 1758, 4to. FRANCIS MASERES. *A Dissertation on the use of the Negative Sign in Algebra.*

London, 1796, 8vo. WILLIAM FREND. *The Principles of Algebra.**

Cambridge, 1803, 4to. ROBERT WOODHOUSE. *The Principles of Analytical Calculation.*

Philosophical Transactions for 1806. M. L'ABBÉ BUÉE. *Mémoire sur les Quantités Imaginaires* (Read June 20, 1805). See also the review of this in Vol. XII. of the *Edinburgh Review*, April —July, 1808 (written by Playfair).

London, 1817, 4to. BENJAMIN GOMPERTZ. *The Principles and Application of Imaginary Quantities*, Book I., *to which are added some observations on porisms.....*

London, 1818, 4to. BENJAMIN GOMPERTZ. *The Principles and Application of Imaginary Quantities*, Book II., *derived from a particular case of functional projections.....*

Paris, 1828, 8vo. (small). C. V. MOUREY. *La vraie théorie des Quantités Négatives, et des Quantités Prétendues Imaginaires. Dédié aux amis de l'évidence.*

Cambridge, 1828, 8vo. JOHN WARREN. *A Treatise on the Geometrical Representation of the Square Roots of Negative Quantities.*

Philosophical Transactions for 1829. JOHN THOMAS GRAVES. 'An attempt to rectify the inaccuracy of some logarithmic formulæ.' (Read December 18, 1828.)

Philosophical Transactions for 1829. JOHN WARREN. 'Consideration of the objections raised against the geometrical representation of the square roots of negative quantities. (Read February 19, 1829.) The same volume contains JOHN WARREN. 'On the geometrical representation of the powers of quantities, whose indices involve the square roots of negative quantities.' (Read June 4, 1829.)

Cambridge, 1830, 8vo. GEORGE PEACOCK. *A Treatise on Algebra.*

Cambridge, 1837, 8vo. Anonymous [OSBORNE REYNOLDS]. *Strictures on certain parts of 'Peacock's Algebra,'* by a Graduate.

---

* An opponent not only of imaginary but of negative quantities. Perhaps this work suggested M. Buée's memoir. I have a letter in my possession from M. Buée to Mr. Frend, dated June 21, 1801, by which it appears that the former was desired by a gentleman in whose house he was living (as tutor, perhaps) to write a private reply to Mr. Frend's objections. This letter evidently contains the germs of the views which he afterwards published. See the Annual Report of the Royal Astronomical Society for 1842. According to Dr. Peacock, M. Buée is the first formal maintainer of the geometrical signification of $\sqrt{-1}$.

Philosophical Transactions for 1831. DAVIES GILBERT. 'On the nature of negative and of imaginary quantities.' (Read November 18, 1830.)

London, 1834, 8vo. *Report of the Third Meeting of the British Association for the Advancement of Science.* This volume contains George Peacock 'Report on certain branches of analysis,' a most valuable historical discussion on, among other things, the advance of algebra. I cite from it the following works, which I have either not seen, or cannot immediately obtain. Paris, 1806, ARGAND, *Essai sur la manière de représenter les Quantités Imaginaires dans les constructions géométriques.* Also papers or observations by FRANÇOIS, ARGAND, SERVOIS, GERGONNE, in the *Annales des Mathématiques* for 1813 (and I suppose the following year). Also a paper on the arithmetic of impossible quantities, by PLAYFAIR, in the *Philosophical Transactions for* 1778; with a Reply, by WOODHOUSE, in the same work for 1802, entitled 'On the necessary truth of certain conclusions obtained by aid of imaginary expressions.'

London, 1836, 8vo. Anonymous [GEORGE PEACOCK]. *A Syllabus of a Course of Lectures upon Trigonometry, and the Application of Algebra to Geometry.*

London, 1837, 8vo. A. DE MORGAN. *Elements of Algebra.* 2nd edition.

London, 1837, 8vo. A. DE MORGAN. *Elements of Trigonometry and Trigonometrical Analysis, preliminary to the Differential Calculus,....*

Edinburgh Philosophical Transactions, Vol. XIV. Part 1. D[UNCAN] F[ORBES] GREGORY. 'On the real nature of Symbolical Algebra.' (Read May 7, 1838).

Ladies' Diary. London, 1839, 8vo. (small). THOMAS WHITE. 'On the algebraical expansion of quantity,.... and on the symbol $\sqrt{-1}$, which is usually considered* to denote impossible or imaginary quantity,' (at page 59).

Cambridge Philosophical Transactions, Vol. VII. Part 2. A. DE MORGAN. 'On the Foundation of Algebra. (Read Dec. 9, 1839).

Cambridge Philosophical Transactions, Vol. VII. Part 3. A. DE MORGAN. 'On the Foundation of Algebra, No. II. (Read Nov. 29, 1841).

Paris, 1841, 8vo. M. F. VALLÈS. *Etudes Philosophiques sur la science du calcul.* Première Partie. No more yet published.

Cambridge Philosophical Transactions, Vol. VIII. Part 2. A. DE MORGAN. 'On the Foundation of Algebra, No. III. (Read Nov. 27, 1843).

Cambridge, 1842 & 1845, 8vo. GEORGE PEACOCK. *A Treatise on Algebra.* Vol. I.—Arithmetical Algebra. Vol. II.—Symbolical Algebra and its applications to the geometry of position.

London, 1843, 12mo. MARTIN OHM [translated by ALEXANDER JOHN ELLIS]. *The Spirit of Mathematical Analysis, and its relation to a logical system.*

---

* The author supposes it to be indeterminate, because it can be expanded by help of a divergent series. The paper is marked 'received April 1816.'

TABLE OF CONTENTS.

*** THE REFERENCES ARE TO THE PAGES OF THE WORK.

BOOK I.—TRIGONOMETRY.

CHAPTER I.

Preliminary Notions.

Definition of trigonometry, 1; undulating magnitude, 1; periodic magnitude, 2; suggested by angular magnitude, 2; *grădual* measurement of angle, 3; factors of 360, 3; circumference of circle, 4; π, 4; multiplication and division by π, 5; arc $\div$ radius, 5; *arcual* measurement of angle, 5, 6; gradual and arcual comparisons, 6; gradual measurement of arc, 6.

CHAPTER II.

On the Trigonometrical Functions, and on Formulæ of One Angle.

Axes, origin, projections, co-ordinates, abscissa, ordinate, 7; r, θ, x, y, 7; sign of r, x, y, 8; four quarters and their signs, 8; θ and $2m\pi + \theta$, 9; sine, cosine, tangent, cotangent, secant, cosecant, versed sine, coversed sine, 9; complement, supplement, opponent, completion, 10; trigonometrical functions as abstract numbers and multipliers, 10; curve of sines, &c., 11; fundamental equations, 11, 12; limits of value, 12; signs, 12, 13; negative sign of r, 13; initial and terminal values, 13; cosine even, sine odd, 13; tangent odd, 14; $\frac{1}{2}m\pi \pm \theta$ and its rules, 14, 15; double value of functions, 16; 15°, 18°, 30°, 45°, 60°, 72°, 75°, 16, 17; $\sin\theta \div \theta$, $(1 - \cos\theta) \div \theta$, $\tan\theta \div \theta$, 17, 18; θ and $1 - \frac{1}{2}\theta^2$, 18; older system of definitions, 18, 19; areæ of circle and sector, 20.

CHAPTER III.

Formulæ which involve two or more Angles.

Extended notion of projection, 21; distinction of AB and BA, and consequences, 21; similar distinction as to angles, 21; signs of projections, 22; general investigation of $\cos(\phi \pm \theta)$ and $\sin(\phi \pm \theta)$, 23; connexion of the formulæ, 24; cases of limited demonstration, 25, 26; collection of formulæ, 26, 27; remarks on the formulæ, 28, 29; cosine and sine of the sum of any number of angles, 29, 30; $\cos n\theta$ and $\sin n\theta$, 30, 31; trisection of an angle, 31; series for $\cos\theta$ and $\sin\theta$, 32, 33; ditto for $\tan\theta$, 34; algebraic definition of trigonometry, 34; $\cos^n\theta$ and $\sin^n\theta$, 34, 35, 36.

CHAPTER IV.

On the Inverse Trigonometrical Functions.

Functional notation, direct and inverse, 37; inverse trigonometrical functions, 37, 38; examples in the use of inverse symbols, 38, 39, 40.

CHAPTER V.

Introduction of the unexplained Symbol $\sqrt{-1}$.

Remarks on the evidence of $\sqrt{-1}$ in *this* chapter, 41; connecting formulæ of trigonometricals and exponentials, 42; De Moivre's theorem, 42; multiplicity of directions in $\theta \div n$, 43, 44; roots, particularly of unity, 45, 46; transformations of $a + b\sqrt{-1}$, and selection of meaning in $\tan^{-1}(b \div a)$, 46; extension of logarithms with Naperian base, 47; extension of the Naperian base, 48; isolated case of coincidence of logarithms in different systems, 48; the negative quantities which have real logarithms, 49; equivalents of De Moivre, 49; deduction of ordinary formulæ from them, 49, 50; reduction of $\sin^m\theta \cos^n\theta$ to a linear form, 50, 51; mode of finding $\Sigma a_n \cos n\theta\, x^n$ and $\Sigma a_n \sin n\theta\, x^n$, 51, 52; connexion of $\phi(x+y\sqrt{-1})$ and $\phi(x-y\sqrt{-1})$, 52, 53; examples, 53, 54, 55; series for $\tan^{-1}x$, 55; calculation of π, 55, 56; inverse connexion of trigonometricals and exponentials, 57; use of multiplicity of value of logarithms, 57, 58; resolution of $\sin\theta$ into factors, 58, 59, 60; Wallis's form of $\frac{1}{2}\pi$, 61: deduction of approximate form for $1.2.3...n$, 61, 62; factors of $\cos\theta$; logarithms of $\sin\theta$, $\cos\theta$, $\tan\theta$, 63; Vieta's expression for $\frac{1}{2}\pi$, 63; Bernoulli's numbers, 64; series for $\tan x$, $\cot x$, and $(\varepsilon^x - 1)^{-1}$, 65.

CHAPTER VI.

On the Connexion of Common and Hyperbolic Trigonometry.

Hyberbola, 66; its areas, 67, 68; formation of hyperbolic trigonometry, and connexion of its formulæ with those of ordinary trigonometry, 69, 70.

CHAPTER VII.

On the Trigonometrical Tables.

Arrangement and extent of the tables, 71, 72; distinction of real and tabular logarithms, 72; tables recommended, 73; argument, interval, function, difference, interpolation, 73; method of interpolating, 74; choice of functions for accuracy, 75; tangents of angles near to 90°, 75, 76; first notions of the construction of trigonometrical tables, 76, 77, 78.

CHAPTER VIII.

On the Solution of Triangles.

Meaning of solution, 79; formulæ for right-angled triangles, 79, 80; cases of ditto ditto, 80; tabulated example, 80; formulæ for oblique triangles, 81, 82, 83; cases of ditto ditto, 84, 85; mode of entrance of double solution, 85; tabulated example, 86, 87; mention of occasional rules, 87; reduction of triangular formulæ to identities, 88.

BOOK II.—DOUBLE ALGEBRA.

CHAPTER I.

Description of a Symbolic Calculus.

Object, 89; peculiar symbols, meanings, rules of operation, 89; possible deficiencies of either, 90, 91; complete absence of either, 91, 92; *symbolic* calculus, what, 92; recovery of meaning, *significant* calculus, 93; slight example, with illustration of defects, 93, 94; possibility of more than one mode of restoration to significance, 94; step from specific to universal arithmetic, and thence to ordinary algebra, 95, 96; necessity for other than numerical distinction, and mode in which distinction is suggested by algebra, 95, 96; *single* algebra, phrase whence derived, 96, 97; twofold use of its signs, directive and conjunctive, 97; remarks on the progress of algebra, 98, 99, 100.

CHAPTER II.

On Symbolic Algebra.

Abandonment of meaning, 101; collection of the symbolic laws of algebra, 101, 102, 103; instances of symbolic deduction, 104, 105; reservation of small italic letters to signify combinations of unit symbols, 105.

CHAPTER III.

On Areas and Solids.

Meanings of the fundamental symbols under which algebra becomes a legitimate mode of establishing the second book of Euclid, 106, 107; comparison of an inexplicable symbol of this algebra with one of ordinary algebra, 107, 108; this system not altogether new, 108.

CHAPTER IV.

Preliminary Remarks on Double Algebra.

Suggestion on $\sqrt{-1}$ which gave rise to it, 109; the old algebra to be incorporated, 109; examination of magnitude *of one dimension*, time, gain and loss, 109, 110; a wider basis of significance in length affected by direction, 110, 111; geometrical introduction extension, not restriction, 111, 112; mode of making the application to problems of one dimension, 112, 113, 114; separation of subject-matter and operative direction in arithmetical addition and multiplication, 115, 116.

CHAPTER V.

Signification of Symbols in Double Algebra.

Reason of the term *double algebra*, drawn from the meaning of an isolated symbol, 117; meaning of =, 117; origin, 117; unit-line, 118; axes of length and direction, 118; meaning of $A + B$, $A - B$, $-B$, 118; coincidences of common and extended addition, &c., 118; addition really junction and its result joint effect, 118; meaning of $A \times B$ and $A \div B$, 119; symbolic representation of double-meaning symbol, (a, α), 119; coincidences of common and extended multiplication, &c., 120; roots and powers, without reference to exponents, 120; developed expression of an algebraical theorem, 121; construction of all double symbols by single ones and $\sqrt{-1}$, 122; re-introduction of trigonometry, 122; demonstration of the symbolic rules, 123, 124, 125; deduction of fundamental trigonometrical formulæ, 126; proof of the validity of that deduction, and of its coincidence with ordinary proofs, 127, 128.

CHAPTER VI.

On the Exponential Symbol.

Assumption of arithmetical logarithms, definition of *logometer*, 129; choice of logometer, 129; multiple values of logometers, 130; definition of A^B, 130; limitation of ε, 130; logometric equations, 131; proof of symbolic rules, and limitations, 131; proof of $\varepsilon^{\theta\sqrt{-1}} = \cos\theta + \sin\theta \,.\, \sqrt{-1}$, 132; connexion of ε and the arcual unit, 132, 133, 134; transformation of R^s, 134; example of reduction to significance, 135; fallacy exposed, 136; formulæ which supply those of p. 131 when the limitations are removed, 136, 137.

CHAPTER VII.

Miscellaneous Remarks and Applications.

Extension of logarithms, 138; forms of $\phi(a \pm b\sqrt{-1})$, 138, 139; extension of trigonometrical terms, 139, 140; discontinuity of passage from $\pm$ to $\mp$ in single algebra, 140; interpretation of a problem impossible in single algebra, 140, 141; Cauchy's theorem on the limits of imaginary roots of equations, 141, 142, 143, 144; Paradox of single algebra which disappears in double algebra, 144, 145; change of geometrical representation in passing from real to imaginary, 145, 146.

CHAPTER VIII.

On the Roots of Unity.

Power of considering $(\pm 1)^n$ as either quantitative or directive, 147, 148; properties of the roots of $+1$, 148, 149, 150, 151, 152, 153, 154; solution of $x^n = 1$ when n is prime and $x^{n-1} = 1$ has been solved, 155, 156, 157; Gauss's accession to Euclid's geometry, 157; properties of the roots of -1, 158; formation of recurring expressions, 158, 159; Thomas Simpson's method of section of series, 159, 160.

CHAPTER IX.

Scalar View of Algebraical Symbols.

Law of ascent of algebraical operations, 161; illustrative notation, 162; scalar function, 162; ascent of algebraical operations, 162; imperfection of ordinary notation for scalar representation, 163; other law which this notation follows, 164; general law of scalar ascent, 164; limitation of the scalar function, 164; inverse operation, 165; fundamental basis of algebra, 166; scalar notation in extension of that of ordinary algebra, 166, 167.

ERRATUM.

Page 138, after $\frac{\lambda X}{\lambda B}$ insert $=$.

ERRATA.

[(, *from the top*;), *from the bottom*]

Page	22	line	(14	for	BB	read	BB'
,,	28	,,	3)	,,	prevent	,,	present.
,,	32	,,	1)	,,	$\frac{n\theta-\theta}{3}$	,,	$\frac{n\theta-2\theta}{3}$
,,	32	,,	1)	,,	$m+1$	,,	$m-1$
,,	33	,,	(1	,,	$m+1$	,,	$m-1$
,,	48	,,	(11	,,	value	,,	values
,,	54	,,	(4	,,	$\tan {}^2\alpha$	,,	α^2
,,	63	,,	6)	,,	—	,,	$\div$
,,	121	,,	(3	,,	*unity*	,,	-1
,,	131	,,	(14	,,	$A^\wedge$	,,	A^1
,,	131	,,	7)	,,	$=A^B$	,,	$=, A^{BC}$
,,	138	,,	11)	after	$\frac{\lambda X}{\lambda B}$	insert	$=$
,,	143	,,	(15	for	cos	read	sin
,,	143	,,	4)	,,	cos	,,	sin
,,	144	,,	(19	,,	0	,,	1
,,	145	,,	12)	,,	limit	,,	unit
,,	149	,,	(7	,,	than l	,,	than n
,,	149	,,	4)	,,	whence	,,	when
,,	150	,,	10)	,,	1, 2, 7	,,	1, 5, 7.

In the List which follows the preface, the printer has omitted a note of interrogation which followed the word *Playfair*.

BOOK I.

TRIGONOMETRY.

CHAPTER I.

PRELIMINARY NOTIONS.

It is proved in the sixth book of Euclid that when sides and diagonals are given, in number enough to determine a rectilinear figure, *angles* depend solely upon the *proportions* of sides, and *proportions* of sides solely upon *angles*. If two angles of a triangle be given, all the ratios of sides are given: and, If the ratio of each of two sides to a third be given, all the angles are given. There is then a close connexion between *angles*, and *ratios of lines:* the branch of mathematics in which this connexion is examined, suitable modes of expression invented, and results obtained and applied, is called TRIGONOMETRY, taking its name from one of its earliest applications, the *measurement of triangles.*

Trigonometry contains *the science of continually undulating magnitude;* meaning magnitude which becomes alternately greater and less, without any termination to succession of increase and decrease. A function of x is continually undulating, when, as x increases continuously, say from 0 to ∞, ϕx never becomes permanently increasing nor permanently diminishing, nor permanently approaching to a fixed limit. Ordinary algebra has no such functions in its finite forms; and though it has them in its infinite series, yet it cannot easily recognize and establish the undulating property. Trigonometry is the *branch of algebra* in which undulating functions are considered. All trigonometrical functions are not undulating: but it may be stated that in common algebra nothing but infinite series undulate: in trigonometry nothing but infinite series do not undulate.

Trigonometry is a *branch of algebra:* nevertheless, it is usually founded on geometrical considerations. This is not an absolute necessity: but any other foundation would make it much more difficult for the beginner to understand. It will become evident that another mode of establishing the algebra of undulating quantities might have been chosen, in which no geometrical notion need have been even alluded to.

Of all *undulating* magnitudes, the most simple is the *periodic*, which exhibits a perpetual recurrence of the same cycles of alteration. If, however a function may change while x changes from 0 to a, the same changes take place while x changes from a to $2a$, from $2a$ to $3a$, and so on, that function is periodic. The general property of such a function is expressed by the equation $\phi(x+a)=\phi x$, true for all values of x, and for one value of a, or *for its multiples*. For $\phi(x+a+a)=\phi(x+a)=\phi x$; or $\phi(x+2a)=\phi x$. Similarly $\phi(x+3a)=\phi x$; and so on.

Consideration of angular magnitude must suggest periodic functions. Let a straight line, fixed at one extremity, revolve about that extremity. The total angle described may go on increasing *ad infinitum:* the angle itself is not a periodic magnitude, though beginners are apt to think so. But the direction indicated is periodic, though not a magnitude.

There is no direction indicated during the second revolution, which was not previously indicated during the first. Let a wheel turn on an horizontal axle, by a handle at the end of a spoke: the angle turned through by the spoke goes on increasing as long as the wheel turns one way; but the height of the handle above the ground is a periodic magnitude, which goes through the same cycle of changes in each and every revolution. Certain periodic functions, suggested by the revolution of a straight line about a point, form the trigonometrical alphabet, as we shall see.

If we had now to invent a mode of measuring angles, the most convenient first method would be to adopt the whole revolution* as a standard unit; thus the angle ·467 would signify

* Observe this consequence of the *periodic character of direction*, that the angle has a unit *expressible in words* without reference to other magnitude *exhibited*. 'The angle through which a line revolves in regaining the direction with which it first started,' is a perfect description of a definite amount of angular magnitude. But no number of volumes could describe an English foot, if drawing, and reference to length supposed capable of access, were both prohibited.

467-1000ths of a revolution. This plan has not been adopted: the usual method is to divide the whole revolution into 360 equal parts, each of which is called a *degree*. Each degree is divided into 60 *minutes*, each minute into 60 *seconds;* formerly each second was divided into 60 *thirds*, each third into 60 *fourths*, and so on. This *sexagesimal* mode of division was once applied to all kinds of magnitude: thus the sixtieth part of *any* length, time, weight, area, &c. was called *its* minute, the sixtieth part of the minute *its* second, and so on. Nothing of this method remains to us, except in the divisions of a degree of angle, a degree of arc (the 360th part of the whole circumference of a circle), and the hour of time. Thus it would have been said that the circumference of a circle is very near to $3\ 8'\ 29''\ 44'''$ diameters, meaning $3+\frac{8}{60}+\frac{29}{3600}+\frac{44}{216000}$ of a diameter.

Degrees, minutes, seconds, &c. are represented by $^{\circ}\ '\ ''\ '''$ &c. But it must be noticed that *thirds*, *fourths*, &c. are wholly obsolete, decimal fractions of the second being preferred. Thus $18^{\circ}\ 47'\ 23''\cdot 1774$ indicates the following fraction of a whole revolution,

$$\frac{18}{360}+\frac{47}{60\times 360}+\frac{23}{60\times 60\times 360}+\frac{1774}{10000}\times\frac{1}{60\times 60\times 360}.$$

In this mode of measurement it is worth while to remember;—the right angle and its multiples, 90°, 180°, 270°, 360°; the half of a right angle and its multiples, 45°, 90°, 135°, 180°, 225°, 270°, 315°, 360°; and the third of a right angle and its multiples, 30°, 60°, 90°, 120°, 150°, 180°, 210°, 240°, 270°, 300°, 330°, 360°. Also the thirds of the revolution, 120°, 240°, 360°; and its fifths, 72°, 144°, 216°, 288°, 360°. And 360 should be well known as $2^3\times 3^2\times 5$, from which its separation into pairs of factors, 2.180, 3.120, 4.90, 5.72, 6.60, 8.45, 9.40, 10.36, 12.30, 15.24, 18.20, will be easily gathered.

The above method of measurement may be called *grădual* (pronounced *grade-ual*). But it is not the only method in use. There is another, which I shall call the *arcual** method. To explain this method, it must first be shewn that circumferences

* I have been in the habit of styling this the *theoretical* method, as being used in the theory of the subject: but I shall now adopt the term used in the text.

of circles are to one another as their diameters. Let it be granted that the circumference of a circle is greater than that of any inscribed polygon, and less than that of any circumscribed polygon.

Draw the circle whose radius is OA. Let BOA be the $2n$th part of a revolution: so that $2n\times BD$ and $2n\times CA$ are the circumferences of the inscribed and circumscribed *regular* polygons of n sides. These circumferences are as BD to CA, or as OB to OC. Consequently, if the angle BOA be made small enough, or n great enough, the inscribed and circumscribed regular circumferences may be made as nearly equal as we please; and either, therefore, as near as we please to the circumference of the circle, which lies between them in magnitude. Now take another figure like the preceding, but constructed on a different radius $O'B'$, and with all its letters accented. We know then that the two inscribed regular circumferences of n sides are to one another as $O'B'$ to OB; and also the two circumscribed circumferences. Let P and P', C and C', Q and Q' be the circumferences of the inscribed polygons, the circles, and the circumscribed polygons. Then the order of magnitude is always P, C, Q and P', C', Q', and the ratios $P:P'$ and $Q:Q'$ are always equal and constant (each being the ratio of the radii) while P and Q, and also P' and Q', can be made as nearly equal as we please. Hence it follows that $C:C'$ is the same ratio as $P:P'$ and $Q:Q'$. Let P and Q be $C-M$ and $C+N$; and let P' and Q' be $C'-M'$ and $C'+N'$. Then M, N, M', N', may each be made as small as we please; and $C-M:C'-M'$ being always one ratio (that of the radii), the limiting ratio $C:C'$ can be no other (Algebra, p. 157). The same follows from the same use of the ratio $C+N:C'+N'$.

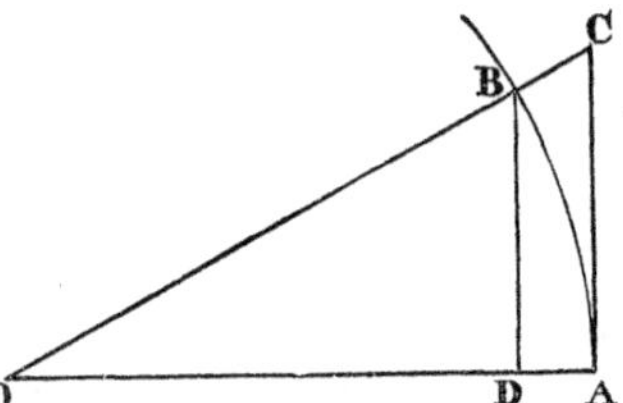

The circumference being C, and the radius R, it follows that the fraction $C\div R$ is the same for all circles. It is always denoted by 2π; that is, π is always made to represent the fraction which expresses the ratio of the circumference to the *diameter*. An investigation of the value of π, such as we can hereafter make, but of which at present we must assume the result, shews that it is nearly $\frac{22}{7}$, very much nearer to $\frac{355}{113}$, and expressed, as far as twenty places of decimals will do it, by 3·14159265358979323846. Its reciprocal is, to the same extent, ·31830988618379067153. I leave the student to demonstrate the following rules, the convenience of which is the formation of results by successive corrections, so that the point at which it is desirable to stop is pointed out by the value of the corrections.

To multiply by π, first take the multiplicand 3 times and one-seventh of a time, deduct its 800th part, the 100th part of the last, and 2 millionths of the multiplicand. Then add the hundred-millionth of the multiplicand, and $7\frac{1}{3}$ per cent. of that hundred-millionth. The result is as correct as if thirteen figures had been used in the ordinary multiplication.

To divide by π, take seven 22nds of the dividend, one 8000th, 3 millionths, and 7 hundred-millionths; then deduct 2 thousand-millionths, and add the thousandth of the last. The result is as correct as if thirteen decimals had been used in the ordinary division.

Let there be an angle of which the arc is s to the radius r, and s' to the radius r', the circumferences being c and c'. Then the angle is to four right angles (Euc. VI. 33) as s to c, and as s' to c'. Hence $s : 2\pi r :: s' : 2\pi r'$, whence $\frac{s}{r} = \frac{s'}{r'}$. Or, to a given subtending angle, arcs are to one another as their radii. Let there be another angle, having the arcs S and S' to the radii r and r'. Then the angles are as s and S, or as $\frac{s}{r}$ and $\frac{S}{r}$, or as $\frac{s'}{r'}$ and $\frac{S}{r}$. That is, any two angles being made central angles in any two circles, the fractions obtained from $\frac{\text{arc}}{\text{rad.}}$ are proportional to the two angles. For instance, the angle which has an arc 6 to the radius 17 is to the angle which has an arc 11 to the radius 8 as $\frac{6}{17}$ to $\frac{11}{8}$.

From this theorem is derived the *arcual* mode of measuring angles. Let the *arcual* angular unit be that angle which subtends an arc equal to the radius, and let all other angles be measured by the numbers of arcual units, or the fractions of an arcual unit, which they contain. Then we shall have the following theorem: The number of arcual units in any angle is the quotient of any arc which that angle subtends, divided by the radius. For if θ be the number of arcual units in the angle which subtends s to the radius r, we have (Euc. VI. 33),

$$\theta : 1 :: s : r, \quad \text{or } \theta = \frac{s}{r}.$$

When we write the equation angle $= \dfrac{\text{arc}}{\text{rad.}}$, we understand by 'angle' an abbreviation of 'number of arcual units contained in the angle'.

The number of arcual units in four right angles is circumference ÷ radius, or 2π; in two right angles, π; in one right angle, $\frac{1}{2}\pi$. Since 180 degrees make π arcual units, the arcual unit is $\dfrac{180}{\pi}$ degrees, or $57^\circ\cdot295779513$; it is also $3437'\cdot74677$; and $206264''\cdot806$. It may be remembered, within the hundredth part of a second, as 57 degrees and *three* tenths, all but one-*fourth* of a minute and one-*fifth* of a second. This is $57^\circ\ 17'\ 44''\cdot8$.

The degree, minute, and second, are severally the fractions ·01745329, ·0002908882, and ·000004848137, of an arcual unit. The arcual unit being our usual reference, the degree may generally be considered as *a small angle.* Most of the theorems which I assert to be approximately true for small angles, are nearly true for an angle as small as a degree.

The student must remember not to confound 2π with 360, nor π with 180, as is sometimes done, even by writers. That $2\pi = 360$ is true in a certain sense; and so is $20 = 1$, for 20 *shillings* are one *pound.*

When a circle is divided into 360 equal *arcs,* each is called a degree of *arc;* and the degree of arc is divided sexagesimally. The radius is $57^\circ\ 47'\ 44''\cdot8$ *of arc.* On a great circle of the earth (the equator for instance, or a meridian), the *second* of arc is about 100 feet.

CHAPTER II.

ON THE TRIGONOMETRICAL FUNCTIONS, AND ON FORMULÆ OF ONE ANGLE.

LET two straight lines be drawn at right angles to one another; let them be called *axes*, and their point of intersection, O, the *origin*. Let any line, OP, be drawn from the origin; and let

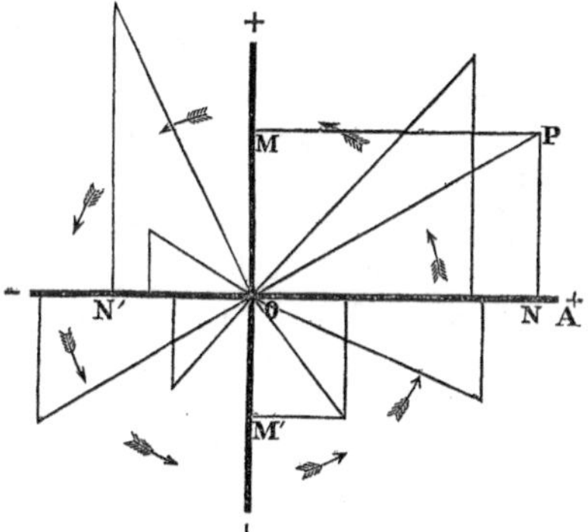

PN, PM be drawn perpendicular to the axes. In the rectangle $MOPN$, ON and OM are called *projections* of OP *upon the axes*. The projections of OP are also called *coordinates* of the point P: and the coordinates are distinguished by the names *abscissa* and *ordinate*. Usually, a projection and a parallel to the other projection are employed, as ON, NP: and then the projection is generally named the *abscissa* of P, and the parallel to the other projection, the *ordinate* of P. And generally the abscissa is taken upon the axis drawn horizontal in the page, and the ordinate parallel to the vertical axis. The letter x usually designates an abscissa, y an ordinate; and the axes are called the axes of x and of y.

A line terminating at O, and indefinitely extended, revolves about O, setting out from one side of the axis of x, OA. When it has described an angle θ, which may be of any magnitude, a distance r is taken off. This distance is always considered as positive, when r is taken off on the revolving line: and as negative, if taken off on the opposite side. Thus, the acute angle POA being 30°, we may refer OP to a line which makes 30° with OA, and then we say that OP is positive. But when we say that OP is part of a line which makes 210° with OA, we call it negative.

Again, one particular direction of revolution is considered as positive, the other as negative. If the arrows designate the positive revolution, then OP, being positive, makes an angle with OA, which may be called $+30°$ or $-330°$; but if OP be negative, it makes an angle $+210°$ or $-150°$.

On the axes, each species of coordinate or projection has its proper algebraical sign. The starting-line of revolution is always taken as the positive side of the axis of x; and the result of $+90°$ of revolution as that of the axis of y. Thus ON is positive, ON' is negative;* OM is positive, OM' is negative.

The axes divide the plane into four quarters: and as a line, revolving positively, passes from 0 to 90°, from 90° to 180, from 180° to 270°, and from 270° to 360°, it is said to be in the first, second, third, and fourth quarters of space. But these might equally well be designated as the $++$, $+-$, $--$, and $-+$ quarters of space.

In this system, $++$, $+-$, $--$, $-+$, the first of each pair gives the succession $++--$; and these are the signs of the y projections of lines in the four quarters: the projection on the axis of y of a line in the first quarter of space, is $+$; in the second, $+$; in the third, $-$; in the fourth, $-$. The second of each pair gives the succession $+--+$; and these are the signs of the x projections of lines in the four quarters. The algebraical combination of each of the pairs gives the succession $+-+-$; and these are the signs

---

* When the revolving line comes into the position ON', *is it negative?* I answer, no: ON', as a projection, is considered as part of a line which makes an angle 0° with the starting-line; and, on a line *so described*, is negative. But ON', as a position of the *line of revolution*, is part of a line which makes 180° with the starting-line; and thus considered, it is positive. The same considerations apply to the other axis. A line may be considered as making *with itself* an angle 0° or an angle 180°: whatever signs its parts have in the first case, they have the opposite ones in the second.

of the arithmetical products or quotients derived from the two projections of a line in each of the four quarters.

Everything that takes place in the first revolution is repeated in the second; and is repeated in an inverted order in the first negative revolution. In all that depends upon the direction in which an amount of revolution terminates, an addition or subtraction of a whole revolution makes no difference whatever. But in all that depends upon the actual magnitude of the angle revolved through, an alteration by a whole revolution makes an effective difference. Measuring arcually, $2m\pi+\theta$ may *most often* be confounded with θ when m is any integer, positive or negative; but not *always*.

The *primary trigonometrical functions* of an angle are the ratios of the projections to the revolving line, and to one another, direct and inverse: these ratios are independent of the length of the revolving line. Let x, y, r be the values, *with their proper signs*, of the abscissa, ordinate, and radius, or base, perpendicular, and hypothenuse. The six ratios $\frac{x}{r}$, $\frac{y}{r}$, $\frac{y}{x}$, $\frac{x}{y}$, $\frac{r}{x}$, $\frac{r}{y}$ take each a name, the etymology of which cannot be explained till we come to exhibit the older definitions: at present they must stand for arbitrary sounds. Let θ be the angle by revolving through which r has gained its position.

			is called the	abbreviated into
$\frac{x}{r}$	$\frac{\text{abscissa}}{\text{rad.}}$	$\frac{\text{base}}{\text{hyp.}}$	*cosine* of θ	$\cos\theta$
$\frac{y}{r}$	$\frac{\text{ordinate}}{\text{rad.}}$	$\frac{\text{perpend.}}{\text{hyp.}}$	*sine* of θ	$\sin\theta$
$\frac{y}{x}$	$\frac{\text{ordinate}}{\text{abscissa}}$	$\frac{\text{perpend.}}{\text{base}}$	*tangent* of θ	$\tan\theta$
$\frac{x}{y}$	$\frac{\text{abscissa}}{\text{ordinate}}$	$\frac{\text{base}}{\text{perpend.}}$	*cotangent* of θ	$\cot\theta$
$\frac{r}{x}$	$\frac{\text{rad.}}{\text{abscissa}}$	$\frac{\text{hyp.}}{\text{base}}$	*secant* of θ	$\sec\theta$
$\frac{r}{y}$	$\frac{\text{rad.}}{\text{ordinate}}$	$\frac{\text{hyp.}}{\text{perpend.}}$	*cosecant* of θ	$\text{cosec}\,\theta$
	$1-\cos\theta$		*versed sine* of θ	$\text{vers}\,\theta$
	$1-\sin\theta$		*coversed sine* of θ	$\text{covers}\,\theta$

This table must be thoroughly learned. The terms base, perpendicular, and hypothenuse, referring to the right-angled triangle in which the projections are sides, does not mean that what Euclid would call *an angle* of that triangle is always the angle in question. It is so when θ is less than a right angle, or when the revolving line is in the first quarter. But in the second quarter, θ is a *supplement** of Euclid's angle; in the third quarter it is an *opponent;* in the fourth quarter it is a *completion.* All this, and many other things of which only hints are given, must be fixed in the mind by attentive consideration of all the phases of the figure of a line projected on the axes: no amount of description will supply the place of such consideration.

It is important to remember that *all the trigonometrical functions are purely abstract numbers.* They are not angles, nor lines, any more than they are weights, or sums of money. They represent the fractions which lines are of lines, the ratios of lines to lines. Thus, the cosine of $60°$ is $\frac{1}{2}$: *one-half of what?* Answer, *one-half of a time:* when the revolving line has described $60°$, the projection on the axis of x is *one-half* of the revolving line; the last words in italics contain the assertion that $\cos 60° = \frac{1}{2}$.

Thus the functions may be advantageously remembered by *their effect as multipliers.* The cosine and sine may be called *projecting factors:* multiplication by $\cos\theta$ turns r into its projection on the axis of x; multiplication by $\sin\theta$ turns r into its projection on the axis of y. The projections of r are $r\cos\theta$ and $r\sin\theta$. The tangent and cotangent are *interchanging factors;* multiplication by $\tan\theta$ converts the projection on x into that on y, multiplication by $\cot\theta$ converts the projection on y into that on x.

We may of course take a line which has as many linear units as a certain angle has of angular units, or as a sine or tangent has

---

* The term *supplement* has long been used to signify the defect from two right angles: thus θ and $\pi - \theta$ are supplements. By *opponents,* I mean angles made by opposite straight lines with one straight line, in the same direction of revolution: thus θ and $\pi + \theta$ are opponents. By *completions,* I mean angles which together make up a whole revolution: thus of θ and $2\pi - \theta$ each is the *completion* of the other. Finally, the well-known term *complement* is arbitrarily used to denote the defect from a right angle: thus θ and $\frac{1}{2}\pi - \theta$ are complements.

of abstract units, and in this sense it may be permitted (to those who can do it without confusion) to talk of a line and angle being equal, or of a line equal to the sine of an angle. The frontispiece has curves constructed in this manner for each of the six principal functions. The origin is O, the axis of x is OA...: the abscissa is the angle, the ordinate on one curve is the sine, &c. The student may, when he has read a little further, detect for himself the curve of sines, of cosines, of tangents, of cotangents, of secants, of cosecants.

There are eight trigonometrical functions, of which two are absolutely *defined* by formulæ; namely,

$$\text{vers}\,\theta = 1 - \cos\theta, \quad \text{covers}\,\theta = 1 - \sin\theta.$$

Of the remaining six, we may predict that five independent equations exist among them: for one angle and one ratio of sides absolutely determine all the angles (and therefore all the ratios of sides) of a triangle, *whensoever that given angle is a right angle or more.* There are easily found more than five relations; but not all independent. First, there are the relations which obviously and necessarily follow from the *algebraical form* of the definitions, independently of the meaning of the symbols. These are

$$\begin{aligned} &\cos\theta \times \sec\theta = 1, \\ &\sin\theta \times \text{cosec}\,\theta = 1, \quad \tan\theta = \frac{\sin\theta}{\cos\theta}, \quad \cot\theta = \frac{\cos\theta}{\sin\theta}. \\ &\tan\theta \times \cot\theta = 1. \end{aligned}$$

Of these only four are independent: the third and fourth make the fifth follow. Secondly, there are the relations which follow from the meaning of x, y, and r. The equation $x^2 + y^2 = r^2$, which follows from the application of arithmetic to Euc. I. 47, gives

$$\left(\frac{x}{r}\right)^2 + \left(\frac{y}{r}\right)^2 = 1, \quad 1 + \left(\frac{y}{x}\right)^2 = \left(\frac{r}{x}\right)^2, \quad 1 + \left(\frac{x}{y}\right)^2 = \left(\frac{r}{y}\right)^2,$$

$$\cos^2\theta + \sin^2\theta = 1, \quad 1 + \tan^2\theta = \sec^2\theta, \quad 1 + \cot^2\theta = \text{cosec}^2\theta,$$

of which one only is independent; for $\cos^2\theta + \sin^2\theta = 1$, gives

$$1 + \left(\frac{\sin\theta}{\cos\theta}\right)^2 = \left(\frac{1}{\cos\theta}\right)^2, \quad \text{or } 1 + \tan^2\theta = \sec^2\theta, \text{ \&c.}$$

The following collection of formulæ, either proved above, or easily deduced, should be carefully remembered:

$$\cos\theta.\ \sec\theta = 1,\quad \cos^2\theta + \sin^2\theta = 1,\qquad \tan\theta = \frac{\sin\theta}{\cos\theta},$$

$$\sin\theta.\operatorname{cosec}\theta = 1,\qquad 1 + \tan^2\theta = \sec^2\theta,\qquad \cot\theta = \frac{\cos\theta}{\sin\theta},$$

$$\tan\theta.\ \cot\theta = 1,\qquad 1 + \cot^2\theta = \operatorname{cosec}^2\theta$$

$$\cos\theta = \frac{1}{\sqrt{(1+\tan^2\theta)}},\quad \sin\theta = \frac{\tan\theta}{\sqrt{(1+\tan^2\theta)}}.$$

$$\text{If } \tan\theta = \frac{b}{a},\quad \cos\theta = \frac{a}{\sqrt{(a^2+b^2)}},\quad \sin\theta = \frac{b}{\sqrt{(a^2+b^2)}}.$$

The student should, as an exercise, express each function in terms of all the rest.

I now proceed to the examination of several material points connected with the several functions.

1. *Limits of value.* No cosine nor sine can fall *without* the interval $-1\ldots+1$: for neither x nor y can numerically exceed r. For the same reason, no secant nor cosecant can fall *within* the interval $-1\ldots+1$. But a tangent or cotangent may have any value, positive or negative. Versed and coversed sines always lie in the interval $0\ldots2$.

2. *Signs.* Let r be taken positively. Then $\frac{x}{r}$ and $\frac{y}{r}$ have the signs of x and y. Hence $\cos\theta$ (and its reciprocal $\sec\theta$) have their signs remembered by the succession $+--+$: or the cosine has the sign $+$ in the first quarter, $-$ in the second, &c. But $\sin\theta$ (and its reciprocal $\operatorname{cosec}\theta$) have their signs remembered by the succession $++--$. And $\tan\theta$ (and its reciprocal $\cot\theta$) have their signs remembered by the succession $+-+-$ (page 8).

The pairs of signs by which the quarters are distinguished, state the signs of the sine and cosine, while the sign compounded of the other two, states that of the tangent. Thus, in the fourth, or $-+$ quarter, the sine is negative, the cosine positive, and the tangent negative.

It is also worth while to remember, that while all the three pairs are positive in the first quarter, each of the other quarters has only one positive pair belonging to it.

In the first quarter, all are positive.
In the second .... only the the sine and cosecant.
In the third .... only the tangent and cotangent.
In the fourth .... only the cosine and secant.

The whole system remains consistent with itself if negative values of r be introduced, under the definitions in page 8. Take the figure in page 7 as an instance, and say that OP is negative. Then, ON and NP being positive, the sine and cosine are negative. And so they ought to be: for OP, being negative, must be considered as on a line in the third quarter, and θ as between two and three right angles.

Versed sines and coversed sines are always positive.

3. *Initial or terminal values.* These are the values when the revolving line begins or ends a quarter of space, and is on one of the axes. In every such case one of the projections vanishes, and the other is of the same length (but not always of the same sign) as the revolving line itself. At $\theta = 0$, and $\theta = \pi$, y vanishes; at $\theta = \frac{1}{2}\pi$, and $\theta = \frac{3}{2}\pi$, x vanishes. At $\theta = 2\pi$, the values at $\theta = 0$ are repeated. Examination will give the following table, which should be remembered; partly by the help of the connecting equations.

Arcual Angle	0	$\frac{1}{2}\pi$	π	$\frac{3}{2}\pi$	2π
Cosine	1	0	-1	0	1
Sine	0	1	0	-1	0
Tangent	0	∞	0	∞	0
Cotangent	∞	0	∞	0	∞
Secant	1	∞	-1	∞	1
Cosecant	∞	1	∞	-1	∞
Versed sine	0	1	2	1	0
Coversed sine	1	0	1	2	1
Gradual Angle	$0°$	$90°$	$180°$	$270°$	$360°$

It is hardly necessary to say that all the trigonometrical functions are periodic, and that 2π is in every case the angular extent of *one* period, or of a number of periods. In every case $F\theta = F(\theta + 2\pi)$, F representing a primary trigonometrical function. The period of the sine and cosine is 2π; of the tangent, π. I now proceed to examine circumstances connected with this periodic character.

The *cosine* is what is called an *even function* of θ; that is, it does not change at all when θ is changed into $-\theta$: or $\cos(-\theta) = \cos\theta$. But the *sine* is what is called an *odd function;* that is, it changes sign when θ is changed into $-\theta$; or $\sin(-\theta) = -\sin\theta$. Let two equal lines revolve, one positively and one negatively: it is clear from the elements of geometry, that whatever equal angles they may have described, the projections on x are the same, identically, and the projections on y differ in sign only. Hence, $x \div r$ is the same for both; and $y \div r$ is not, but the difference is in sign only.

The *tangent* is an *odd* function; for $\tan(-\theta) = \sin(-\theta) \div \cos(-\theta)$ $= -\sin\theta \div \cos\theta = -\tan\theta$. The cotangent is also an odd function. The secant and versed sine are even functions; the cosecant is odd; the coversed sine is neither The terms even and odd, as applied to functions in general, are suggested by the properties of the *even* and *odd* powers.

If, θ and the length of the revolving line being given, we form a new angle thus, one or more right angles $\pm\theta$, it will readily be seen that the right-angled triangle made by the revolving line is in all cases the same in form and magnitude. But two variations of position occur; sometimes the projections differ in sign from those of the original triangle; sometimes they change name, the line which was x becoming y, and *vice versâ*. An examination of all the cases will present the following table:—

Angle.	Absc.	Ordin.	Conclusions.
θ	x	y	
$\frac{1}{2}\pi-\theta$	y	x	$\cos(\frac{1}{2}\pi-\theta)=\sin\theta$, $\sin(\frac{1}{2}\pi-\theta)=\cos\theta$, $\tan(\frac{1}{2}\pi-\theta)=\cot\theta$,
$\frac{1}{2}\pi+\theta$	$-y$	x	$\cos(\frac{1}{2}\pi+\theta)=-\sin\theta$, $\sin(\frac{1}{2}\pi+\theta)=\cos\theta$, $\tan(\frac{1}{2}\pi+\theta)=-\cot\theta$,
$\pi-\theta$	$-x$	y	$\cos(\pi-\theta)=-\cos\theta$, $\sin(\pi-\theta)=\sin\theta$, $\tan(\pi-\theta)=-\tan\theta$,
$\pi+\theta$	$-x$	$-y$	$\cos(\pi+\theta)=-\cos\theta$, $\sin(\pi+\theta)=-\sin\theta$, $\tan(\pi+\theta)=\tan\theta$,
$\frac{3}{2}\pi-\theta$	$-y$	$-x$	$\cos(\frac{3}{2}\pi-\theta)=-\sin\theta$, $\sin(\frac{3}{2}\pi-\theta)=-\cos\theta$, $\tan(\frac{3}{2}\pi-\theta)=\cot\theta$,
$\frac{3}{2}\pi+\theta$	y	$-x$	$\cos(\frac{3}{2}\pi+\theta)=\sin\theta$, $\sin(\frac{3}{2}\pi+\theta)=-\cos\theta$, $\tan(\frac{3}{2}\pi+\theta)=-\cot\theta$,
$2\pi-\theta$	x	$-y$	$\cos(2\pi-\theta)=\cos\theta$, $\sin(2\pi-\theta)=-\sin\theta$, $\tan(2\pi-\theta)=-\tan\theta$.

These transformations, it must be observed, apply to *all* values of θ. For instance, let θ lie between $\frac{3}{2}\pi$ and 2π; then $\frac{1}{2}\pi-\theta$ lies between $-\pi$ and $-\frac{3}{2}\pi$. Draw the figure accordingly, and it will appear that the x of either is the y of the other, *both in sign and magnitude.* The formulæ therefore are universally true:

but they may be best remembered by the supposition that θ is a small angle, so that $\frac{1}{2}\pi - \theta$ is in the first right angle, $\frac{1}{2}\pi + \theta$

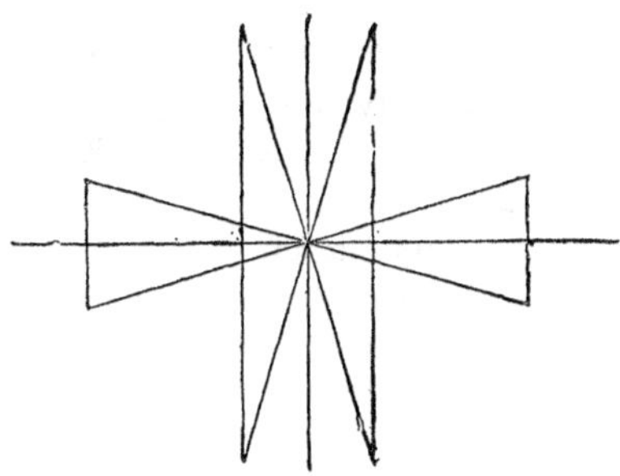

in the second, as also $\pi - \theta$, and so on. All the cases may now be contained in the following rule. According as the number of right angles is even or odd, let the function remain, or let it be changed into its co-function (sine and cosine, &c. are co-functions). Then prefix the sign which the given function has when θ is less than a right angle; and lastly, write θ for the angle. For example, let it be required to simplify $\tan(\frac{3}{2}\pi + \theta)$. There is an odd number of right angles; $\frac{3}{2}\pi + \theta$ is in the fourth right angle, when $\theta < \frac{1}{2}\pi$: in the fourth right angle the tangent is negative: accordingly, $\tan(\frac{3}{2}\pi + \theta) = -\cot\theta$. But in transforming $\cot(\pi - \theta)$, we see an even number of right angles, and an angle in the second right angle; accordingly, $\cot(\pi - \theta) = -\cot\theta$.

The following cases are so important that they should be remembered apart:—

The functions of complements are co-functions,

$$\sin(\tfrac{1}{2}\pi - \theta) = \cos\theta, \quad \cos(\tfrac{1}{2}\pi - \theta) = \sin\theta,$$

Supplements have the same sine, $\sin(\pi - \theta) = \sin\theta$.

Opponents have the same tangent, $\tan(\pi + \theta) = \tan\theta$.

Completions have the same cosine, $\cos(2\pi - \theta) = \cos\theta$,

$$\cos(\tfrac{1}{2}\pi + \theta) = -\sin\theta, \quad \cos(\pi - \theta) = -\cos\theta,$$

$$\sin(\tfrac{1}{2}\pi + \theta) = \cos\theta, \quad \tan(\pi - \theta) = -\tan\theta.$$

All the angles which have the same sine as θ are included in the formulæ $2m\pi + \theta$, and $(2m + 1)\pi - \theta$: all which have the same cosine as θ in the formulæ $2m\pi + \theta$ and $2m\pi - \theta$: all which have the same tangent as θ in $m\pi + \theta$: m being any integer, positive or negative.

To one sine there is but one cosecant; to one cosine there is but one secant; to one tangent there is but one cotangent; and *vice versâ*. But in every other case a function has two functions of every other kind attached to it, with opposite signs. This appears, firstly, from what precedes: any sine, for instance, belongs to two angles, supplements, which have cosines, &c. opposite in sign. Supplements, opponents, and completions have their functions of equal value, *and opposite signs*, except in the three cases noted above. Thus,

$$\sin(\pi + \theta) = -\sin\theta, \quad \tan(2\pi - \theta) = -\tan\theta, \text{ \&c.}$$

Secondly, from the equations in page 12, in which it appears that every determination of one function in terms of *one* other function requires an extraction of the square root, except when the functions are reciprocals. Thus,

$$\cos\theta = \pm\sqrt{(1 - \sin^2\theta)}, \quad \tan\theta = \pm\sqrt{(1 - \cos^2\theta)} \div \cos\theta, \quad \text{\&c.}$$

When one function is found, all the others can be found. And by very ordinary geometry we can contrive to express the functions of 15°, 18°, 30°, 45°, 60°, 72°, 75°. Of these I shall deduce some, and arrange the whole in a table, of which I leave the student to fill up the demonstration.

(45°). This angle has equal projections, or $\tan 45° = 1$.

(30°, 60°). If a, a, c be the sides of an isosceles triangle, and 2θ its vertical angle, then $c = 2a\sin\theta$. Let the triangle be equilateral; then $a = 2a\sin 30°$; or $\sin 30° = \frac{1}{2}$. And this is $\cos 60°$.

(18°, 72°). By Euc. IV. 10, it appears that an isosceles triangle having 36° for its vertical angle has for its base the greater segment of the side, as determined in II. 11. If then a be the side, and c the base, we have $a(a-c) = c^2$, or $2c = (\sqrt{5} - 1)a$. Hence, $\sin 18°$ (or $\cos 72°$) $= \frac{1}{4}(\sqrt{5} - 1)$.

(15°, 75°). Take a right-angled triangle having an angle of 30°, an hypothenuse 2, and therefore 1 for the side opposite and $\sqrt{3}$ for the side adjacent to that angle. Bisect the angle of 30°; the bisecting line divides 1 into segments which are as $\sqrt{3}$ to 2: the smaller segment is then $\sqrt{3} \div (2 + \sqrt{3})$, or $\sqrt{3}(2 - \sqrt{3})$, or $2\sqrt{3} - 3$.

The bisecting line is therefore

$$\sqrt{(3+12+9-12\sqrt{3})}, \text{ or } \sqrt{12}.\sqrt{(2-\sqrt{3})}, \text{ or } 2\sqrt{3}.\frac{\sqrt{3}-1}{\sqrt{2}}.$$

Hence, $\sin 15^\circ = \dfrac{\sqrt{3}(2-\sqrt{3})}{\sqrt{3}(\sqrt{6}-\sqrt{2})} = \dfrac{\sqrt{3}-1}{2\sqrt{2}} = \dfrac{\sqrt{6}-\sqrt{2}}{4}.$

		sine	cosine	tangent	cotangent		
$\frac{1}{12}\pi$	15°	$\frac{\sqrt{6}-\sqrt{2}}{4}$	$\frac{\sqrt{6}+\sqrt{2}}{4}$	$2-\sqrt{3}$	$2+\sqrt{3}$	75°	$\frac{5}{12}\pi$
$\frac{1}{10}\pi$	18°	$\frac{\sqrt{5}-1}{4}$	$\frac{\sqrt{(10+2\sqrt{5})}}{4}$	$\frac{\sqrt{5}-1}{\sqrt{(10+2\sqrt{5})}}$	$\frac{\sqrt{(10+2\sqrt{5})}}{\sqrt{5}-1}$	72°	$\frac{2}{5}\pi$
$\frac{1}{6}\pi$	30°	$\frac{1}{2}$	$\frac{1}{2}\sqrt{3}$	$\frac{1}{3}\sqrt{3}$	$\sqrt{3}$	60°	$\frac{1}{3}\pi$
$\frac{1}{4}\pi$	45°	$\frac{1}{2}\sqrt{2}$	$\frac{1}{2}\sqrt{2}$	1	1	45°	$\frac{1}{4}\pi$
		cosine	sine	cotangent	tangent		

Let θ be arcually measured, and let it be the $2n$th part of four right angles. Describe a circle with the revolving line r for its radius, and inscribe a regular polygon of n sides. One of these sides is $2r\sin\theta$, and the whole circumference of the polygon is $2nr\sin\theta$; that of the circle is $2\pi r$, or $2n\theta.r$. The ratio of these circumferences is $\dfrac{\sin\theta}{\theta}$; and this ratio, the larger n is made, or the smaller θ, the more near it is to unity (page 4). That is, when θ diminishes without limit, the fraction $\sin\theta \div \theta$ approaches without limit to unity. The approach is distinctly seen, even when the angle is far from very small, to ordinary notions: 5°, in arcual units, is ·0872665, and its sine is ·0871557: these two differ by less than the 800th part of either.

Again, $\dfrac{\tan\theta}{\theta} = \dfrac{1}{\cos\theta}\cdot\dfrac{\sin\theta}{\theta}$; $\dfrac{1-\cos\theta}{\theta} = \dfrac{\theta}{1+\cos\theta}\cdot\left(\dfrac{\sin\theta}{\theta}\right)^2$;

the second of which is easily got from $1-\cos^2\theta = \sin^2\theta$. It follows that the limit of $\tan\theta \div \theta$, as θ diminishes without limit, is 1×1 or 1; while that of $(1-\cos\theta)\div\theta$ is 0×1 or 0. As θ diminishes, then, $\sin\theta$ and $\tan\theta$ approach to θ, but $1-\cos\theta$ diminishes much more rapidly than θ. Any number being named, however great, θ contains $1-\cos\theta$ more than that number of times, before θ becomes nothing. When θ is ·0872665 (5° of gradual measure), θ contains $1-\cos\theta$ more than 20 times.

It is important to observe that $\frac{\sin\alpha\theta}{\sin\beta\theta}$, or $\frac{\sin\alpha\theta}{\alpha\theta}\cdot\frac{\beta\theta}{\sin\beta\theta}\cdot\frac{\alpha}{\beta}$, has $\frac{\alpha}{\beta}$ for its limit, when θ diminishes without limit. Also that $n\sin\frac{\theta}{n}$ has the limit θ, when n increases without limit.

We have seen that $\tan\theta$ and $\sec\theta$ both become infinite when $\theta=\frac{1}{2}\pi$. Now

$$\sec\theta-\tan\theta=\frac{1-\sin\theta}{\cos\theta}=\frac{\cos\theta}{1+\sin\theta}\ (=0,\text{ when }\theta=\tfrac{1}{2}\pi);$$

consequently, as θ approaches $\frac{1}{2}\pi$, the difference of $\tan\theta$ and $\sec\theta$ diminishes without limit. Shew in a similar way that $\operatorname{cosec}\theta-\cot\theta$ diminishes without limit, with θ.

$$\text{Again, } 1-\cos\theta=\frac{\sin^2\theta}{1+\cos\theta}\ (=\frac{\theta^2}{1+1}\text{ nearly, when }\theta\text{ is small}).$$

Hence, when θ is small, $\cos\theta=1-\frac{1}{2}\theta^2$ nearly. This, and $\sin\theta=\theta$, are equations which are near enough to truth for most purposes of calculation, when θ is small.

I now give an account of the method of defining the trigonometrical terms which was,* until very lately, universal.

A given straight line, called the *radius*, revolves from a starting-line OA, as in our definitions; but it must be of the same length for all angles, which need not be the case in ours. The *arc* described by the revolving extremity generally (though not always) takes the place of the angle.† Then BM was called the *sine* of the *arc* AB (*sinus*, bosom, the literal translation of an Arabic word: if BAB''' represent a bow (*arcus*), half of the string BB''' comes against the breast of the archer). And OM is the *cosine* of AB: this word is an abbreviation of sine of the complement, or *complemental sine;* it was long before OM was considered as anything but the sine of another arc, BA'. And AM (once the *sagitta*, as occupying the place of the *arrow*) was the *versed sine* (or *turned* sine) of the arc AB. And $A'L$ should have been called the *coversed-sine*, as being the versed sine of the complement: but this term is only a recent invention for the completion

* But not from all time; for Rheticus, who gave the first complete trigonometrical table, and invented the secant and cosecant to complete it, used the method of ratios.

† By constant attention to the arc of a circle, some writers have become unable to think of *angle* as a *magnitude*.

of the system. Draw a *tangent* at A, and at A'; and produce

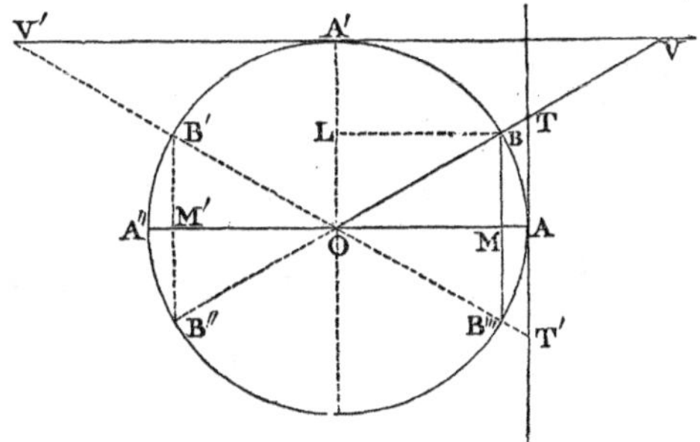

OB to meet them in T and V. Then AT was called the *tangent* (as being drawn on the tangent) of AB; and $A'V$, the tangent of the complement, was called the cotangent of AB. Lastly, OT was called the *secant* of AB, as being on a line which cuts the circle; and OV the *cosecant.*

All these definitions are thus connected with ours: the *old linear function,* divided by *the radius,* in every case gives the *modern numerical function.*

Denote the linear function by the word commencing with a capital letter; and let $OB = r$, $\angle BOA = \theta$. Then we have

$$\sin\theta = \frac{MB}{OB} = \frac{\text{Sin}\,AB}{r}; \quad \tan\theta = \frac{AT}{OA} = \frac{\text{Tan}\,AB}{r};$$

$$\cos\theta = \frac{OM}{OB} = \frac{\text{Cos}\,AB}{r}; \quad \sec\theta = \frac{OT}{OA} = \frac{\text{Sec}\,AB}{r};$$

$$\cot\theta = \frac{OM}{MB} = \frac{A'V}{OA'} = \frac{\text{Cot}\,AB}{r}; \quad \text{vers}\,\theta = 1 - \frac{OM}{OB} = \frac{AM}{OB} = \frac{\text{Vers}\,AB}{r};$$

$$\text{cosec}\,\theta = \frac{OB}{MB} = \frac{OV}{OA'} = \frac{\text{Cosec}\,AB}{r}; \quad \text{Covers}\,\theta = 1 - \frac{MB}{OB} = \frac{A'L}{OB} = \frac{\text{Covers}\,AB}{r}.$$

Speaking but of ultimate calculation, the old system is identical with the new one, if we only make $r = 1$, or take the linear unit for a radius. But there always remains this essential distinction, that the function of the old system is *always* a line, that of the new one a *number.* In the old system the sine of the *arc* of 30° is *half a radius,* whether that radius be used as the measuring unit* or not. In the modern system, the sine of the *angle* of 30° is *the fraction which half a radius is of the whole radius.*

If we substitute in our formulæ the equivalents of the old system, we have such equations as

$$\frac{\text{Sin}\,AB}{r} \cdot \frac{\text{Cosec}\,AB}{r} = 1, \quad \text{or} \quad \text{Sin}\,AB \cdot \text{Cosec}\,AB = r^2,$$

$$\frac{\text{Cos}^2 AB}{r^2} + \frac{\text{Sin}^2 AB}{r^2} = 1, \quad \text{or} \quad \text{Cos}^2 AB + \text{Sin}^2 AB = r^2, \ \&\text{c}.$$

If it be occasionally desirable to refer to the old system, it may be done without confusion by speaking of the *sines*, &c. of *arcs* or the *linear sines*, &c. of angles.

The area of the circle is thus found. Inscribing the polygon of n sides, and θ being the $2n^{\text{th}}$ part of a revolution, or $\frac{\pi}{n}$, we have for each of the n triangles, the area $\frac{1}{2}r\cos\theta \cdot 2r\sin\theta$, and for the whole polygon $nr^2\cos\theta\sin\theta$, or $r^2\cos\frac{\pi}{n} \cdot n\sin\frac{\pi}{n}$. When n increases without limit, the limit of this is πr^2, which is the area of the whole circle. The sector which has the angle θ, being to the whole circle as θ to 2π, is $\frac{\theta r^2}{2}$.

---

* Trigonometry might be defined as that part of the application of Algebra to Geometry which is *independent of linear measure;* since ratios are independent of the units in which their terms are arithmetically expressed. One disadvantage of the old system is, that it keeps this independence of linear measure out of view.

CHAPTER III.

FORMULÆ WHICH INVOLVE TWO OR MORE ANGLES.

It will be desirable to gain a more extended notion of the projections of a line. If any line, AB, be taken as belonging to an indefinite line on which sign is recognized, a careful distinction must be drawn between AB and BA: one is positive and the other is negative. Thus, p. 7, NO is not x, but ON, which is there positive, while NO is negative. If we attend to this, we shall find that, however A, B, C, may be distributed on a straight line,

$$AC = AB + BC = AB - CB = BC - BA; \quad AB + BC + CA = 0;$$

$-$ A C B $+$

Thus $AC = +2$, $AB = +7$, $BC = -5$, and $+2 = +7 + (-5)$.

Next, the angle made by P with Q is to be carefully distinguished from the angle made by Q with P. If one be θ the other is $-\theta$, or, at our pleasure, $2\pi - \theta$. To gain fixed ideas, let us suppose that in the angle made by P with Q, denoted by $P \wedge Q$, we proceed from the positive direction of Q as a starting-line, and thence by revolution, positive or negative according as we want a positive or negative angle, to the positive direction of P. But in $Q \wedge P$, we proceed from the positive direction of P to that of Q. Ascertain from this that

$$P \wedge Q + Q \wedge P \text{ is } 2\pi, 0, \text{ or } -2\pi,$$

according as we take the positive angles in both cases, or one positive and one negative, or both negative.

When neither end of a line is at the origin, the projections are determined by drawing perpendiculars from both ends of the line. Thus AA' has NN' and MM' for projections: but

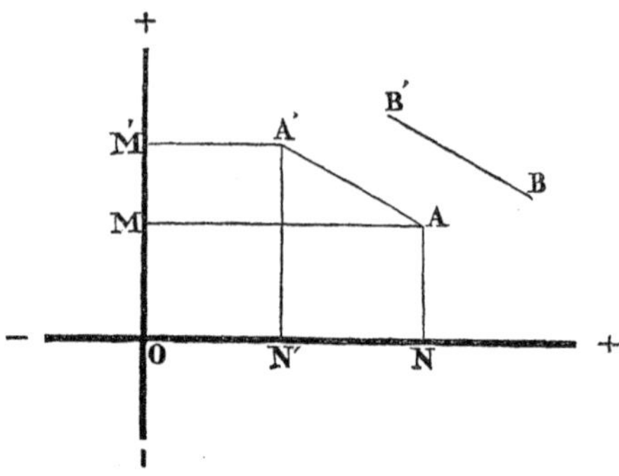

$A'A$ has $N'N$ and $M'M$.

The projections of the line r, making the angle θ with the axis of x, are always $r\cos\theta$ and $r\sin\theta$. Take the preceding figure, and first let AA' be +. If θ be the angle it makes with the axis of x, that angle belongs to the + – quarter: $r\sin\theta$ is, as to sign, $+\times+$, and is positive; and so is MM'. But $r\cos\theta$ is $+\times-$, or negative; and so is NN'. But if AA' be negative, the angle (the opposite side being now used, as in p. 8) is of the – + quarter; and $r\sin\theta$ is $-\times-$, or +, as before; while $r\cos\theta$ is $-\times+$, or –, as before.

In the language of Euclid, equal and parallel lines have equal projections. But *we* must say, equal and parallel lines, *estimated in the same directions*, have equal projections. Thus AA' and BB have equal projections; and so have $A'A$ and $B'B$: but $A'A$ and BB' have only projections equal in length, and differing in sign.

If any points, as A, B, C, D, be taken, the projection of AD is the algebraical sum of those of AB, BC, and CD.

These projections, taking the axis of x, are PS, PQ, QR, RS and, by what precedes,

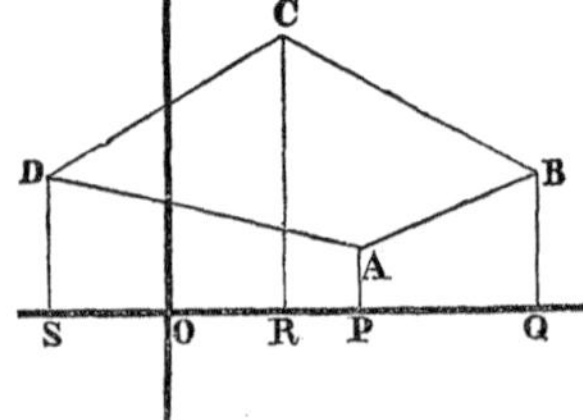

$$PS = PR + RS = PQ + QR + RS.$$

The only question now is this, do PS, PQ, QR, RS, always represent the projections, in what manner soever the lines AD, &c. take signs? And the answer is in the affirmative, since we have seen that any variation of sign is accompanied by a compensating variation in the mode of estimating the angle; so that the projection remains unaltered both in sign and magnitude, so long as the line remains unaltered in direction and magnitude.

Now let the axes revolve through the angle ϕ, giving a pair

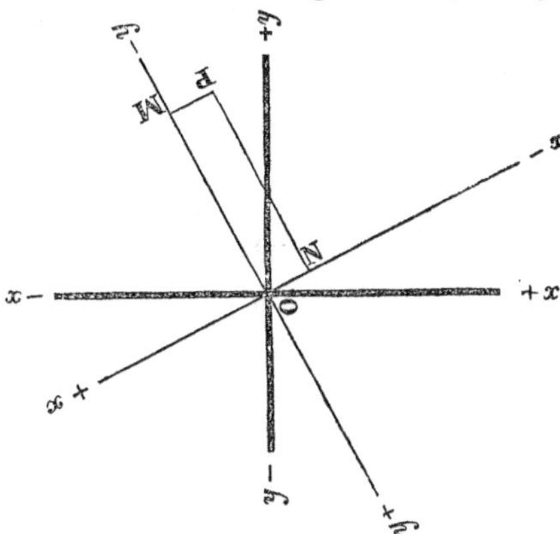

of *secondary* axes; and let a revolving line, starting from the positive side of the secondary axis x, revolve through a further angle θ; having thus revolved through $\phi + \theta$ from the original starting-line. In the diagram, ϕ is about $2\frac{1}{2}$ right angles, and θ is nearly three right angles more; or, if you please, a little more than a right angle negatively. The projections of OP on the primary axes are $r\cos(\phi+\theta)$ and $r\sin(\phi+\theta)$; on the secondary axes, $r\cos\theta$ and $r\sin\theta$; and these last projections, ON, and NP or OM, make angles with the primary axis of x which, estimated by our rules, are ϕ and $\phi+\frac{1}{2}\pi$; for the revolving axis of y is always a right angle in advance of the axis of x. If then we project the secondary projections on the primary axes, we have

Projections of ON are $r\cos\theta.\cos\phi$ and $r\cos\theta.\sin\phi$,

Projections of NP are $r\sin\theta.\cos(\phi+\frac{1}{2}\pi)$ and $r\sin\theta.\sin(\phi+\frac{1}{2}\pi)$.

Looking at the projections on the primary axis of x, we have

Projection of OP = Projection of ON + Projection of NP,

$$r\cos(\phi+\theta) = r\cos\theta\cos\phi + r\sin\theta.\cos(\phi+\tfrac{1}{2}\pi),$$

$$\cos(\phi+\theta) = \cos\theta\cos\phi + \sin\theta\{\cos(\phi+\tfrac{1}{2}\pi) \text{ or } -\sin\phi\},$$

$$\cos(\phi+\theta) = \cos\phi\cos\theta - \sin\phi\sin\theta.$$

Looking at the projections on the primary axis of y, we have

Projection of OP = Projection of ON + Projection of NP,

$$r\sin(\phi+\theta) = r\cos\theta\sin\phi + r\sin\theta\sin(\phi+\tfrac{1}{2}\pi),$$

$$\sin(\phi+\theta) = \cos\theta\sin\phi + \sin\theta\{\sin(\phi+\tfrac{1}{2}\pi) \text{ or } \cos\phi\},$$

$$\sin(\phi+\theta) = \sin\phi\cos\theta + \cos\phi\sin\theta.$$

These formulæ being universally true, we might write $-\theta$ instead of θ, and then we have

$$\cos(\phi-\theta) = \cos\phi\cos(-\theta) - \sin\phi\sin(-\theta),$$
$$= \cos\phi\cos\theta + \sin\phi\sin\theta,$$
$$\sin(\phi-\theta) = \sin\phi\cos(-\theta) + \cos\phi\sin(-\theta),$$
$$= \sin\phi\cos\theta - \cos\phi\sin\theta.$$

This foundation of all the ulterior part of trigonometry, may be stated thus,

$$\cos(\phi\pm\theta) = \cos\phi\cos\theta \mp \sin\phi\sin\theta,$$
$$\sin(\phi\pm\theta) = \sin\phi\cos\theta \pm \cos\phi\sin\theta.$$

The formulæ are not independent: but any one really contains all. This has partially appeared. To shew it completely, observe that the operations connected with projection on the axis of y are precisely the same as those connected with the axis of x. If we adopt the axis of y as a starting-line, and preserve the positive direction of revolution unaltered, we may reckon angles from the axis of y, *and use cosines in determining the projections*, provided that every line which makes an angle μ with the axis of x, be considered as making $\mu-\frac{1}{2}\pi$, or $\mu+\frac{3}{2}\pi$, whichever we please, with the axis of y. If then we want to apply the formula

$$\cos(\phi+\theta) = \cos\phi\cos\theta - \sin\phi\sin\theta,$$

to angles measured from y, we must alter ϕ, which is measured from x, into $\phi-\frac{1}{2}\pi$. This gives

$$\cos(\phi+\theta-\tfrac{1}{2}\pi)[\text{or } \cos\{\tfrac{1}{2}\pi-(\phi+\theta)\}] = \cos(\phi-\tfrac{1}{2}\pi)\cos\theta - \sin(\phi-\tfrac{1}{2}\pi)\sin\theta,$$

or $$\sin(\phi+\theta) = \sin\phi\cos\theta + \cos\phi\sin\theta.$$

The demonstration above given is universal: but it can only be convincing to those who enable themselves to understand, in the most general sense, the preliminary theorems. Any want of such mastery over the *universal* character of theorems in projection will follow the student through all his course, particularly in the higher geometry and in mechanics. To break the difficulty, it may be worth while to examine demonstrations of particular cases, so as to show what manner of *arithmetically* separate operations are algebraically presented in one by the preceding process.

[Within these brackets, lines are not affected by any but *specified* sign: thus AB when negative is written $-AB$: and no distinction is made between AB and BA.

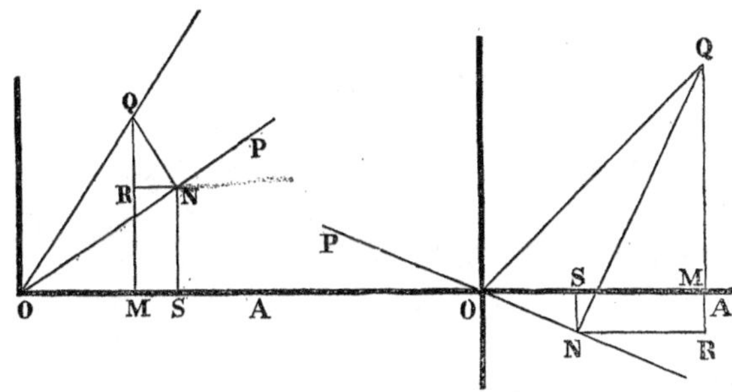

Let $AOP=\phi$, $POQ=\theta$, both taken in the positive direction of revolution. In the second diagram ϕ is nearly two, and θ nearly three, right angles. Project OQ on OA and OP, &c. into OM, MQ and ON, NQ; project ON, NQ into OS, SN, and NR, RQ. In the first diagram, in which ϕ and θ have positive sines and cosines (and $RQN=\phi$), we have

$$OQ\cos(\phi+\theta)=OM=OS-SM=OS-RN,$$
$$=ON\cos\phi-QN\sin\phi=OQ\cos\theta\cos\phi-OQ\sin\theta\sin\phi,$$
$$\cos(\phi+\theta)=\cos\phi\cos\theta-\sin\phi\sin\theta.$$

Also $$OQ\sin(\phi+\theta)=QM=RM+QR=NS+QR,$$
$$=ON\sin\phi+QN\cos\phi=OQ\cos\theta\sin\phi+OQ\sin\theta\cos\phi,$$
$$\sin(\phi+\theta)=\sin\phi\cos\theta+\cos\phi\sin\theta.$$

In the second diagram, RQN is not ϕ, but $\pi-\phi$. And first,

$$OQ\cos(\phi+\theta-2\pi)=OM=OS+SM=OS+RN$$
$$=ON\cos NOS+QN\sin NQR$$
$$=OQ\cos QON\cos NOS+OQ\sin QON\sin NQR;$$
$$\cos(\phi+\theta-2\pi)=\cos QON\cos NOS+\sin QON\sin NQR$$

$$\cos QON = \cos(\theta - \pi) = \cos(\pi - \theta) = -\cos\theta;$$

$$\sin QON = \sin(\theta - \pi) = -\sin(\pi - \theta) = -\sin\theta;$$

$$\cos NOS = \cos(\pi - \phi) = -\cos\phi; \ \sin NQR = \sin(\pi - \phi) = \sin\phi.$$

Whence $\cos(\phi + \theta - 2\pi) = (-\cos\theta)(-\cos\phi) + (-\sin\theta)(\sin\phi);$

or $\cos(\phi + \theta) = \cos\phi\cos\theta - \sin\phi\sin\theta.$

Again,

$$\begin{aligned} OQ\sin(\phi+\theta-2\pi) &= QM = QR - RM = QR - NS \\ &= QN\cos NQR - NO\sin NOS \\ &= OQ\sin QON\cos NQR - OQ\cos QON . \sin NOS, \end{aligned}$$

$$\begin{aligned} \sin(\phi+\theta-2\pi) &= \sin QON\cos NQR - \cos QON . \sin NOS \\ &= (-\sin\theta)\cos(\pi-\phi) - (-\cos\theta)\sin(\pi-\phi), \end{aligned}$$

or $\sin(\phi+\theta) = \sin\theta\cos\phi + \cos\theta\sin\phi.$

The student should repeat the same process on various cases.]

Observe that a complete proof of the cases of $\cos(\phi+\theta)$ and $\sin(\phi+\theta)$ is also one of $\cos(\phi-\theta)$ and $\sin(\phi-\theta)$, independently of the substitution of $-\theta$ for θ. For $\cos(\phi-\theta)$ is $\cos(\phi+2\pi-\theta)$ or $\cos\phi\cos(2\pi-\theta) - \sin\phi\sin(2\pi-\theta)$ or $\cos\phi\cos\theta + \sin\phi\sin\theta$. And similarly for $\sin(\phi-\theta)$.

From the table in p. 17, verify the first row by aid of the third and fourth: find the sines and cosines of 3°, 12°, 27°, 33°, 48°, 63°. From these the sines and cosines of all the multiples of 3° may be easily expressed.

The only form of the preceding theorems which occurs among the fundamental equations is

$$\cos(\theta-\theta) = \cos\theta . \cos\theta + \sin\theta . \sin\theta \text{ or } 1 = \cos^2\theta + \sin^2\theta.$$

A large collection of formulæ may be deduced, as follows:

1. $\cos(\phi+\theta) = \cos\phi\cos\theta - \sin\phi\sin\theta.$
2. $\cos(\phi-\theta) = \cos\phi\cos\theta + \sin\phi\sin\theta.$
3. $\sin(\phi+\theta) = \sin\phi\cos\theta + \cos\phi\sin\theta.$
4. $\sin(\phi-\theta) = \sin\phi\cos\theta - \cos\phi\sin\theta.$

5. $\cos(\phi-\theta) + \cos(\phi+\theta) = 2\cos\phi\cos\theta.$
6. $\cos(\phi-\theta) - \cos(\phi+\theta) = 2\sin\phi\sin\theta.$
7. $\sin(\phi+\theta) + \sin(\phi-\theta) = 2\sin\phi\cos\theta.$
8. $\sin(\phi+\theta) - \sin(\phi-\theta) = 2\cos\phi\sin\theta.$

For ϕ write $\frac{\phi+\theta}{2}$, and for θ write $\frac{\phi-\theta}{2}$;

9. $\cos\theta + \cos\phi = 2\cos\frac{\phi+\theta}{2} \, . \cos\frac{\phi-\theta}{2}$.

10. $\cos\theta - \cos\phi = 2\sin\frac{\phi+\theta}{2} \, . \sin\frac{\phi-\theta}{2}$.

11. $\sin\phi + \sin\theta = 2\sin\frac{\phi+\theta}{2} \, . \cos\frac{\phi-\theta}{2}$.

12. $\sin\phi - \sin\theta = 2\cos\frac{\phi+\theta}{2} \, . \sin\frac{\phi-\theta}{2}$.

13. $\frac{\sin\phi - \sin\theta}{\sin\phi + \sin\theta} = \frac{\tan\frac{1}{2}(\phi-\theta)}{\tan\frac{1}{2}(\phi+\theta)}$, $\frac{\sin\phi + \sin\theta}{\cos\phi + \cos\theta} = \tan\frac{\phi+\theta}{2}$ &c.

14. $\tan\phi + \tan\theta = \frac{\sin(\phi+\theta)}{\cos\phi\cos\theta}$, $\tan\phi - \tan\theta = \frac{\sin(\phi-\theta)}{\cos\phi\cos\theta}$,

$1 + \tan\phi\tan\theta = \frac{\cos(\phi-\theta)}{\cos\phi\cos\theta}$, $1 - \tan\phi\tan\theta = \frac{\cos(\phi+\theta)}{\cos\phi\cos\theta}$.

15. $\tan(\phi \pm \theta) = \frac{\sin(\phi\pm\theta)}{\cos(\phi\pm\theta)} = \frac{\sin\phi\cos\theta \pm \cos\phi\sin\theta}{\cos\phi\cos\theta \mp \sin\phi\sin\theta}$;

divide numerator and denominator by $\cos\phi\cos\theta$, and

$$\tan(\phi+\theta) = \frac{\tan\phi + \tan\theta}{1 - \tan\phi\tan\theta}, \quad \tan(\phi-\theta) = \frac{\tan\phi - \tan\theta}{1 + \tan\phi\tan\theta},$$

which also follow immediately from 14.

16. $\sin 2\theta = 2\sin\theta\cos\theta$, $\sin\theta = 2\sin\frac{\theta}{2}\cos\frac{\theta}{2}$.

17. $\cos 2\theta = \cos^2\theta - \sin^2\theta = 2\cos^2\theta - 1 = 1 - 2\sin^2\theta$.

18. $\cos^2\theta = \frac{1}{2} + \frac{1}{2}\cos 2\theta$, $\sin^2\theta = \frac{1}{2} - \frac{1}{2}\cos 2\theta$.

19. $1 + \cos\theta = 2\cos^2\frac{\theta}{2}$, $1 - \cos\theta = 2\sin^2\frac{\theta}{2}$, $\frac{1-\cos\theta}{1+\cos\theta} = \tan^2\frac{\theta}{2}$.

20. $\tan 2\theta = \frac{2\tan\theta}{1 - \tan^2\theta}$.

21. $\frac{1 - \sin\theta}{1 + \sin\theta} = \tan^2\left(\frac{\pi}{4} - \frac{\theta}{2}\right)$.

22. $\tan\left(\theta - \frac{\pi}{4}\right) = \frac{\tan\theta - 1}{\tan\theta + 1}$.

The following remarks may be made on these formulæ.

5, 6, 7, 8. Remember these formulæ thus:

product of cosines = half cosine of difference + half cosine of sum,
product of sines = half cosine of difference − half cosine of sum,
sin greater × cos less = half sine of sum + half sine of difference,
sin less × cos greater = half sine of sum − half sine of difference.

The universal formulæ are here expressed (the two last, at least) with some arithmetical limitation; by which the one most convenient for arithmetical operation may be selected. Thus at once we learn to write down

$$\sin 5^\circ \cos 18^\circ = \tfrac{1}{2}\sin 23^\circ - \tfrac{1}{2}\sin 13^\circ, \quad \sin 50^\circ \cos 4^\circ = \tfrac{1}{2}\sin 54^\circ + \tfrac{1}{2}\sin 46^\circ.$$

We have thus convenient substitutes for multiplication of sines and cosines by one another; of which much use was made before the invention of logarithms: We can also resolve any product of sines and cosines. Thus

$$\cos a \sin b \sin c = \cos a \{\tfrac{1}{2}\cos(b-c) - \tfrac{1}{2}\cos(b+c)\}$$
$$= \tfrac{1}{4}\{\cos(b-c-a) + \cos(b-c+a)\} - \tfrac{1}{4}\{\cos(b+c-a) + \cos(b+c+a)\}$$
$$= \tfrac{1}{4}\{\cos(b-c-a) + \cos(b-c+a) - \cos(b+c-a) - \cos(b+c+a)\}.$$

Or thus: $\cos a \sin b \sin c = \{\tfrac{1}{2}\sin(b-a) + \tfrac{1}{2}\sin(b+a)\}\sin c$

$$= \tfrac{1}{2}\{\tfrac{1}{2}\cos(b-a-c) - \tfrac{1}{2}\cos(b-a+c)\} + \tfrac{1}{2}\{\tfrac{1}{2}\cos(b+a-c) - \tfrac{1}{2}\cos(b+a+c)\},$$

the same as before.

9, 10, 11, 12. Remember these formulæ thus:

Sum of sines = twice sine of half sum × cosine of half difference.
Difference of sines = twice cosine of half sum × sine of half *direct** diff.
Sum of cosines = twice cosine of half sum × cosine of half diff.
Difference of cosines = twice sine of half sum × sine of half *inverted* diff.

Most write the formula 10 as

$$\cos\phi - \cos\theta = -2\sin\frac{\phi+\theta}{2}\sin\frac{\phi-\theta}{2}.$$

But whichever way it is written, no one will ever be expert in the use of trigonometrical formulæ until $\cos(a-b)$ and $\cos(b-a)$ prevent the instantaneous notion of perfect identity of value and sign: while $\sin(a-b)$ and $\sin(b-a)$ equally suggest sameness of value with difference of sign. Again, it is frequently desirable,

* *Direct*, read in the order of reference; *inverted*, read in the contrary order. When ϕ, θ, are mentioned in that order, $\phi-\theta$ is the direct difference, $\theta-\phi$ the inverted difference.

after observing the effect of an interchange upon one side of an equation, to verify the sameness of the effect on the other side. Thus in 9, interchange of ϕ and θ produce no alteration in the first side: how is it seen that no alteration is produced on the second side? By remembering that $\cos\frac{1}{2}(\theta-\phi)$ and $\cos\frac{1}{2}(\phi-\theta)$ are the same. Again, interchange of ϕ and θ changes the sign of the first side of 10; and of the second also, since $\sin\frac{1}{2}(\theta-\phi)$ and $\sin\frac{1}{2}(\phi-\theta)$ have different signs. A person thoroughly practised in these considerations remembers the *general character* of the formulæ 9–12, and makes the details correct by the habit of satisfying the above conditions.

15. Two angles differ by a right angle; how are their tangents related? If $\phi=\theta+\frac{1}{2}\pi$, $\tan\phi=-\cot\theta$, or $1+\tan\phi\tan\theta=0$. This result, which is often wanted, is best remembered by the denominator in 15: if $\tan(\phi-\theta)$ be infinite, we must have

$$1+\tan\phi\tan\theta=0.$$

Prove the following formula:

$$\tan(\phi+\psi+\theta)=\frac{\tan\phi+\tan\psi+\tan\theta-\tan\phi\,.\tan\psi\,.\tan\theta}{1-\tan\phi\tan\psi-\tan\psi\tan\theta-\tan\theta\tan\phi};$$

from which it follows that the sum of the tangents of the three angles of a triangle is equal to their product. Also the following: If t_1 be the sum of the tangents of a set of angles, t_2, t_3, &c. the sums of the products of every two, every three, &c.; then the tangent of the sum of those angles is $t_1-t_3+t_5-\ldots$ divided by $1-t_2+t_4-\ldots$. This may best be proved by showing that if it be true for any number of angles, it remains true when one more angle is introduced.

If there be any number of angles, and if S_0 be the product of all their cosines, and S_n the sum of all the products which have for factors the *sines* of n of them and the *cosines* of all the rest; then the sine of the sum of those angles is $S_1-S_3+S_5-\ldots$ and the cosine of the sum is $S_0-S_2+S_4-\ldots$.

Suppose this proposition true for any one number of angles, and let S_0, S_1, &c. have the above meaning. Introduce one more angle, having a and b for its cosine and sine, and let T_0 be *now* the product of all the cosines, and T_n the sum of the products in which n are sines and the rest cosines. Now it is clear that

T_0 is S_0a. Next, T_1 consists, first, of all the terms which compose S_0, each multiplied by b, and of those of S_1 each multiplied by a; whence $T_1 = S_0b + S_1a$. And T_2 has all the terms in S_1 each multiplied by b, and all those in S_2 each multiplied by a; whence $T_2 = S_1b + S_2a$. And thus we show that $T_m = S_{m-1}b + S_ma$. But if there be k angles in the first set, T_{k+1} is S_kb, and S_{k+1} does not exist. But the law of connexion $T_{k+1} = S_kb + S_{k+1}a$ still exists if we suppose $S_{k+1} = 0$.

Now if the cosine and sine of the sum of the k angles be $S_0 - S_2 + S_4 - \ldots$ and $S_1 - S_3 + S_5 - \ldots$, then, after introduction of the new angle, the cosine and sine of the sum of the $k+1$ angles are

$$(S_0 - S_2 + S_4 - \ldots)\,a - (S_1 - S_3 + S_5 - \ldots)\,b \text{ or } T_0 - T_2 + T_4 - \ldots,$$
$$(S_1 - S_3 + S_5 - \ldots)\,a + (S_0 - S_2 + S_4 - \ldots)\,b \text{ or } T_1 - T_3 + T_5 - \ldots.$$

If then the theorem be true for k angles, it is true for $k+1$. But it is true for *two* angles; for, ϕ and θ being those angles, S_0 is $\cos\phi\cos\theta$, and S_1 is $\sin\phi\cos\theta + \cos\phi\sin\theta$, and S_2 is $\sin\phi\sin\theta$, S_3 is 0, S_4 is 0, &c. And $\cos(\phi+\theta)$ is $S_0 - S_2 + S_4 - \ldots$, while $\sin(\phi+\theta)$ is $S_1 - S_3 + S_5 - \ldots$. Hence the theorem is true for three angles, hence for four, &c. The beginner had better proceed in one or two cases thus:

$$\begin{aligned}&\cos(\phi+\psi+\theta)\\ &= \cos(\phi+\psi)\cos\theta - \sin(\phi+\psi)\sin\theta\\ &= (\cos\phi\cos\psi - \sin\phi\sin\psi)\cos\theta - (\sin\phi\cos\psi + \cos\phi\sin\psi)\sin\theta\\ &= \cos\phi\cos\psi\cos\theta - (\sin\phi\sin\psi\cos\theta + \sin\phi\sin\theta\cos\psi + \sin\psi\sin\theta\cos\phi)\\ &= S_0 - S_2 + (S_4 = 0) - (S_6 = 0) + \ldots.\end{aligned}$$

If there be n angles, the number of products having m sines is the number of distinct ways in which we can select m out of the n angles, or the number of combinations of m out of n: denote this by m_n; accordingly

$$m_n \text{ stands for } n\,\frac{n-1}{2}\,\frac{n-2}{3}\,\ldots\,\frac{n-m+1}{m}.$$

If all the angles be equal, and each of them be θ, each term of S_m is $\mathrm{c}^{n-m}\mathrm{s}^m$, where c means $\cos\theta$ and s means $\sin\theta$. Accordingly, S_m becomes $m_n\mathrm{c}^{n-m}\mathrm{s}^m$, and we now have

$$\cos n\theta = \mathrm{c}^n - 2_n\mathrm{c}^{n-2}\mathrm{s}^2 + 4_n\mathrm{c}^{n-4}\mathrm{s}^4 - 6_n\mathrm{c}^{n-6}\mathrm{s}^6 + \ldots,$$
$$\sin n\theta = 1_n\mathrm{c}^{n-1}\mathrm{s} - 3_n\mathrm{c}^{n-3}\mathrm{s}^3 + 5_n\mathrm{c}^{n-5}\mathrm{s}^5 - 7_n\mathrm{c}^{n-7}\mathrm{s}^7 + \ldots.$$

Now the development of $(c + s)^n$ is $c^n + 1_n c^{n-1}s + 2_n c^{n-2}s^2 + \ldots$; whence the following theorem: Develope $(c + s)^n$ by the binomial theorem, and put together the odd terms, 1st, 3rd, 5th, &c., and the even terms, 2nd, 4th, 6th, &c.; change the alternate signs in each lot, and the results are $\cos n\theta$ and $\sin n\theta$. Thus we may at once write down

$$\cos 2\theta = c^2 - s^2, \qquad \sin 2\theta = 2cs,$$
$$\cos 3\theta = c^3 - 3cs^2, \qquad \sin 3\theta = 3c^2s - s^3,$$
$$\cos 4\theta = c^4 - 6c^2s^2 + s^4, \qquad \sin 4\theta = 4c^3s - 4cs^3.$$

The beginner should form some of these successively; thus

$$\begin{aligned}\sin(3\theta) = \sin(2\theta + \theta) &= \sin 2\theta \,.\, c + \cos 2\theta \,.\, s\\ &= 2cs \,.\, c + (c^2 - s^2)\, s = 3c^2s - s^3\\ \sin(4\theta) = \sin(3\theta + \theta) &= \sin 3\theta \,.\, c + \cos 3\theta \,.\, s\\ &= (3c^2s - s^3)\, c + (c^3 - 3cs^2)\, s\\ &= 4c^3s - 4cs^3, \quad \text{and so on.}\end{aligned}$$

The question of finding the sine or cosine of the nth part of an angle is now reduced to that of solving an equation of the nth degree. For example, given the sine of an angle, b, it is required to find the sine of its third part. Here

$$b = 3(1 - x^2)\, x - x^3 = 3x - 4x^3,$$

x being the sine of the third part. Hence x is to be found from $4x^3 - 3x + b = 0$.

For example, if the angle be 30°, we have to solve

$$8x^3 - 6x + 1 = 0,$$

which, by Horner's method, has ·173648177867 for one of its roots, approximately; and this root is sin 10°.

There are three roots to this equation, all real: but three distinct problems are attempted, all soluble. For what we really ask, in the equation, is the sine of the third part of the angle whose sine is $\frac{1}{2}$. This last angle may be either 30°, 360° + 30°, 2 × 360° + 30°, 3 × 360° + 30°, 4 × 360° + 30°, &c., or 180° − 30°, 3 × 180° − 30°, 5 × 180° − 30°, &c. Look among the thirds of all these angles, and we find three angles having distinct sines, 10°, 130°, 250°; or 10°, 50°, 250°. And the three values of x are the sines of these three angles.

From the preceding theorem we can, and with tolerable ease, exhibit the algebraical series which $\cos\theta$ and $\sin\theta$ are equivalent to. First, we must ascertain that, x having a fixed value, $\left(\cos\frac{x}{n}\right)^n$ neither diminishes nor increases without limit when n increases without limit. Of this, *à priori*, we must be uncertain, for as n increases, $\cos\frac{x}{n}$ increases towards unity, while the increase of the exponent has a diminishing effect. Between the increase and the diminution, we are unable to say whether $\left(\cos\frac{x}{2n}\right)^{2n}$, for instance, is greater or less than $\left(\cos\frac{x}{n}\right)^n$. But, taking n so great to begin with as that $\frac{x}{n}$ shall be between $-\frac{1}{2}\pi$ and $+\frac{1}{2}\pi$, we easily see by our formulæ that the duplication of n effects an increase. For

$$\cos^2\frac{x}{2n} = \frac{1+\cos\frac{x}{n}}{2} > \frac{\cos\frac{x}{n}+\cos\frac{x}{n}}{2} > \cos\frac{x}{n};$$

or

$$\left(\cos\frac{x}{2n}\right)^{2n} > \left(\cos\frac{x}{n}\right)^n.$$

Beginning then at $\cos x$, the succession $\cos x$, $\left(\cos\frac{x}{2}\right)^2$, $\left(\cos\frac{x}{4}\right)^4$, &c. is a succession of increasing terms, of which no one exceeds unity: for $\left(\cos\frac{x}{n}\right)^n$ cannot exceed unity, unless $\cos\frac{x}{n}$ could be greater than unity. Accordingly, the preceding terms severally approach to some limit: let it be L.

Now take the term which may represent any one of the terms already found in $\cos n\theta$ and $\sin n\theta$; namely,

$m_n c^{n-m} s^m$, or $n\frac{n-1}{2}\frac{n-2}{3}\ldots(m \text{ factors})\ldots\frac{n-m+1}{m}(\cos\theta)^{n-m}(\sin\theta)^m$.

Let $n\theta = z$, *a fixed angle:* but nevertheless n may be as great as we please, provided θ be taken $= z \div n$. And as n increases without limit, θ diminishes without limit. Now take the term preceding, divide it by $(\cos\theta)^n$, and at the same time multiply and divide it by θ, m times. It then becomes

$$n\theta\frac{n\theta-\theta}{2}\frac{n\theta-\theta}{3}\ldots\frac{n\theta-(m+1)\theta}{m}\cdot\frac{1}{\theta^m}\frac{(\sin\theta)^m}{(\cos\theta)^m};$$

or $$z\,\frac{z-\theta}{2},\ \frac{z-2\theta}{3}\ \ldots\ldots\ \frac{z-(m+1)\,\theta}{m}\left(\frac{\tan\theta}{\theta}\right)^{m}.$$

When θ diminishes without limit, this, for every specific value of m, approaches without limit to $z.\frac{z}{2}.\frac{z}{3}\ldots\frac{z}{m}.1^m$, or $\frac{z^m}{2.3\ldots m}$.

Next, after dividing both equations in page 30 by c^n or $\left(\cos\frac{z}{n}\right)^n$, perform the preceding compensatory operations on the several terms, and equate the limits of the sides of the equations (Algebra, page 157). We have then,

$$\frac{\cos z}{L}=1-\frac{z^2}{2}+\frac{z^4}{2.3.4}-\ldots;\quad \frac{\sin z}{L}=z-\frac{z^3}{2.3}+\frac{z^5}{2.3.4.5}-\ldots$$

These series will be found to be convergent (Algebra, page 186); and these equations themselves determine L. For if we make $z=0$, the first gives $\cos 0=L$, or $L=1$: if we divide both sides of the second by z, and diminish z without limit, remembering that $\sin z\div z$ has the limit 1, we also find $L=1$.

Our results then are (z *being an angle in arcual units*),

$$\cos z=1-\frac{z^2}{2}+\frac{z^4}{2.3.4}-\frac{z^6}{2.3.4.5.6}+\frac{z^8}{2.3.4.5.6.7.8}-\ldots\ldots,$$

$$\sin z=z-\frac{z^3}{2.3}+\frac{z^5}{2.3.4.5}-\frac{z^7}{2.3.4.5.6.7}+\frac{z^9}{2.3.4.5.6.7.8.9}-\ldots\ldots,$$

in which we see verification of the preceding assertions—that $\cos z$ is an even function, and $\sin z$ an odd one — that $\sin z=z$ and $\cos z=1-\frac{1}{2}z^2$, nearly, when z is small.

The readiest mode of calculation from these series is by throwing them into the forms

$$\cos z=1-\frac{z^2}{2}\left\{1-\frac{z^2}{3.4}\left\{1-\frac{z^2}{5.6}\left\{1-\frac{z^2}{7.8}\left\{1-\ldots\ldots,\right.\right.\right.\right.$$

$$\sin z=z\left\{1-\frac{z^2}{2.3}\left\{1-\frac{z^2}{4.5}\left\{1-\frac{z^2}{6.7}\left\{1-\frac{z^2}{8.9}\left\{1-\ldots\ldots,\right.\right.\right.\right.\right.$$

where { indicates that the preceding multiplier is a factor of all that follows.

Thus, the calculation of cos 1 (or 57°17′44″·8 in gradual units) is obtained to twelve decimal places (see the property of alternating series, Algebra, page 184) from

$$\cos 1=1-\frac{1}{2}\left\{1-\frac{1}{3.4}\left\{1-\frac{1}{5.6}\left\{1-\frac{1}{7.8}\left\{1-\frac{1}{9.10}\left\{1-\frac{1}{11.12}\left\{1-\frac{1}{13.14}\right\}.\right.\right.\right.\right.\right.$$

Turn $1 \div 13.14$ into a decimal fraction of 13 places, subtract it from unity, divide by 11.12, subtract from unity, &c., keeping 13 places throughout: the final result may be depended on to 12 places.

Those who have mastery enough over algebraical division to divide the series for the sine by that for the cosine, will find

$$\tan z = z + \frac{z^3}{3} + \frac{2z^5}{15} + \frac{17z^7}{315} + \frac{62z^9}{2835} + \ldots\ldots,$$

the law of the terms of which is too complicated for the beginner. Every one, however, should verify on the series, $\cos^2 z + \sin^2 z = 1$, $\cos^2 z - \sin^2 z = \cos 2z$, $2 \sin z \cos z = \sin 2z$.

I said (page 2) that we should soon make it very evident that a purely algebraical basis *might* have been made for trigonometry. If we had chosen to call the preceding functions of z, namely

$$1 - \frac{z^2}{2} + \ldots, \quad z - \frac{z^3}{2.3} + \ldots, \quad z + \frac{z^3}{3} + \ldots,$$

by the names of *cosine*, *sine*, and *tangent* of z, (and their reciprocals *secant*, *cosecant*, and *cotangent*), we might have investigated the properties of these series, and we should *at last* have arrived at all our preceding formulæ of connexion; but with much more difficulty.

I now go to the converse problem, in which it is required to express $\cos^n\theta$ and $\sin^n\theta$ by means of sines or cosines of θ, 2θ, 3θ, &c.

First, let there be n angles a, b, c, d, &c., and proceed as in page 28 with $\cos a \cos b \cos c$. Thus we have

$$\cos a \cos b = \tfrac{1}{2} \cos(a - b) + \tfrac{1}{2} \cos(a + b)$$

$$\cos a \cos b \cos c = \tfrac{1}{4} \cos (a - b - c) + \tfrac{1}{4} \cos (a - b + c) + \tfrac{1}{4} \cos (a + b - c) + \tfrac{1}{4} \cos (a + b + c), \text{ \&c.}$$

The final divisor will be 2^{n-1}, the final number of cosines 2^{n-1}; and, looking at the manner in which the angles enter, we shall see the cosine of every choice out of $+ a \pm b \pm c \pm d \pm \ldots$ In every term change the sign of every letter, which will not alter the value of any one cosine, add the results together and divide by 2, which will leave the whole unaltered, and we shall then have 2^n for a divisor, 2^n for the number of terms, and every variety of $\pm a \pm b \pm c \pm d \pm \ldots$ among the angles. If we now make a, b, c, &c.

all equal to one another and to θ, we shall be able to subdivide all the choices furnished in $\pm\theta\pm\theta\pm\theta\pm\theta\pm\dots$ (n terms) into the following. One case of $n\theta$, taking all +, and one case of $-n\theta$, taking all $-$: 1_n cases of $(n-2)\theta$, with one only taken negative (giving $(n-1)\theta-\theta$), and as many with $-(n-2)\theta$, taking one only positive: 2_n cases of $(n-4)\theta$, taking two only $-$, and as many of $-(n-4)\theta$, taking two only +; and so on. But at the last step there will be a separation between the cases of n even and n odd. If n be even, say $=2k$, there will be at last k_n cases of $\{(n-k)-k\}\theta$, or 0θ, taking k $-$ and k +; the case of k taken + and k taken $-$ not being distinct from the former. But if n be odd, say $=2k+1$, then there are k_n cases of $\{(n-k)-k\}\theta$ or θ, in which k are taken $-$; and as many of $-\theta$, in which k are taken +. Accordingly, $\cos a \cos b \cos c\dots$being now $\cos^n\theta$, we have

$$2^n\cos^n\theta = [\cos n\theta + \cos(-n\theta) + 1_n\{\cos(n-2)\theta + \cos-(n-2)\theta\}$$
$$+ 2_n\{\cos(n-4)\theta+\cos-(n-4)\theta\} + 3_n\{\cos(n-6)\theta+\cos-(n-6)\theta\}$$
$$+ \dots \text{ ending with } k_n\cos 0\theta \text{ if } n=2k, \text{ and with}$$
$$k_n(\cos\theta+\cos-\theta) \text{ if } n=2k+1].$$

Collecting these, by help of $\cos(-a)=\cos a$, we have, for a final form,

$$\cos n\theta = \frac{1}{2^{n-1}}\left\{\cos n\theta + n\cos(n-2)\theta + n\,\frac{n-1}{2}\cos(n-4)\theta + \dots\right\}$$

on the condition that $\cos 0\theta$, when it occurs, is only to take *half* the coefficient indicated by the general law.

The beginner may proceed thus,

$$\cos^2\theta = \tfrac{1}{2} + \tfrac{1}{2}\cos 2\theta \qquad \cos^3\theta = \tfrac{1}{2}\cos\theta + \tfrac{1}{4}(\cos\theta + \cos 3\theta)$$
$$= \tfrac{1}{4}(\cos 3\theta + 3\cos\theta)$$
$$\cos^4\theta = \tfrac{1}{4}(\cos 3\theta\cos\theta + 3\cos^2\theta) = \tfrac{1}{8}(\cos 2\theta + \cos 4\theta + 3 + 3\cos 2\theta)$$
$$= \tfrac{1}{8}(\cos 4\theta + 4\cos 2\theta + \tfrac{6}{2}\cos 0\theta), \text{ \&c.}$$

Now let θ be changed into $\frac{1}{2}\pi-\theta$, or $\cos^n\theta$ into $\sin^n\theta$. If we examine $\cos\left(m\dfrac{\pi}{2}-A\right)$, we begin by rejecting all the fours out of m, as indicative of complete revolutions; and the final form of this term depends on the remainder. Call $4k$ an even even-number, as it is the $2k^{\text{th}}$ even number, $4k+2$ an odd even-number, being the $(2k+1)^{\text{th}}$; $4k+1$ an odd odd-number, is it the $(2k+1)^{\text{th}}$ odd number; and $4k+3$ an even odd-number, being the

$(2k+2)^{\text{th}}$. Then, for even even-numbers, the above is $+\cos A$; for odd odd-numbers, $+\sin A$; for odd even-numbers, $-\cos A$; for even odd-numbers, $-\sin A$. And $m-2$ is of the class of even numbers, or of odd numbers, of which m is not, &c. We have then the four following formulæ:

$$n \text{ even even } \sin^n\theta = \frac{1}{2^{n-1}}\left\{\cos n\theta - n\cos(n-2)\theta + n\frac{n-1}{2}\cos(n-4)\theta - \ldots\right\}$$

$$n \text{ odd odd } \sin^n\theta = \frac{1}{2^{n-1}}\left\{\sin n\theta - n\sin(n-2)\theta + n\frac{n-1}{2}\sin(n-4)\theta - \ldots\right\}$$

$$n \text{ odd even } \sin^n\theta = -\frac{1}{2^{n-1}}\left\{\cos n\theta - n\cos(n-2)\theta + n\frac{n-1}{2}\cos(n-4)\theta - \ldots\right\}$$

$$n \text{ even odd } \sin^n\theta = -\frac{1}{2^{n-1}}\left\{\sin n\theta - n\sin(n-2)\theta + n\frac{n-1}{2}\sin(n-4)\theta - \ldots\right\}$$

Of these the beginner should construct instances, as before. He may also try to prove the following theorems:

$$\sin 2\theta = \frac{2}{\cot\theta + \tan\theta} = \frac{\tan(45^\circ+\theta) - \tan(45^\circ-\theta)}{\tan(45^\circ+\theta) + \tan(45^\circ-\theta)},$$

$$\cos 2\theta = \frac{1}{1+\tan 2\theta\tan\theta} \qquad \tan\theta = \cot\theta - 2\cot 2\theta,$$

$$\frac{1-\cos\theta}{\sin\theta} = \tan\frac{\theta}{2} \qquad \frac{1+\cos\theta}{\sin\theta} = \cot\frac{\theta}{2}.$$

CHAPTER IV.

ON THE INVERSE TRIGONOMETRICAL FUNCTIONS.

WE may now consider $\cos\theta$, &c. as functions of θ, and accordingly, θ itself as a function of $\cos\theta$, or of $\sin\theta$, &c. If $x = \cos\theta$, then θ may be described as 'an angle whose cosine is x.' The continental writers denote this by angle (cosine $= x$), but in our country it is universally described by a symbol derived from a functional analogy. If ϕx denote a function of x, then $\phi(\phi x)$ is denoted by $\phi^2 x$, $\phi(\phi^2 x)$ by $\phi^3 x$, and so on. On this notation $\phi^{-1}x$ should denote the function on which performance of ϕ gives x, so that $\phi(\phi^{-1}x) = x$, and $\phi^{-2}x$ should denote that function which gives $\phi^2(\phi^{-2}x) = x$. Accordingly, $\cos x$ being considered as a function of x, and the abbreviated word *cos* as a functional symbol, $\cos^{-1}x$ should denote the function which satisfies $\cos(\cos^{-1}x) = x$. Hence, $\cos^{-1}x$ must stand for the angle (meaning *any* angle) whose cosine is x. Similarly, $\sin^{-1}x$, $\tan^{-1}x$, &c. stand severally for any angle whose sine, tangent, &c. is x.

The objection to this analogy is, that we do not pursue it. We do not employ $\sin^2 x$ for *the sine* of the sine* of x, but for $\sin x \times \sin x$ or $(\sin x)^2$. The answer is, that we *ought* to follow the analogy, and that we certainly should, if questions in which the sine of the sine, or the sine of the sine of the sine, were so frequently employed as to require abbreviation. And when such questions actually occur, then $\sin^3 x$ should stand for $\sin\{\sin(\sin x)\}$, and $\sin x \times \sin x \times \sin x$ should be denoted by $(\sin x)^3$; a form which some writers prefer, as it is. But as sines of sines, &c. *very* rarely occur, it is not necessary to disturb established notation.

As above defined, ϕx and $\phi^{-1}x$ are what are called inverse functions. But when we talk of *the* inverse function of ϕx, it

* The student may ask, How can any thing but *an angle* have a sine? I answer, that θ is not an *angle*, but the *number* of arcual units in an angle. Every *number* has a *sine*.

is as when we talk of *the* square root of x, knowing that there are *two*. Generally speaking, inverse functions have more values than one; in the case before us, an infinite number. For, let $\cos^{-1}x$ be θ, that is, let θ be an angle which has x for its cosine; then so has $2m\pi \pm \theta$, m being any integer, positive or negative.

Hence it always arises that though $\phi\phi^{-1}x = x$, $\phi^{-1}\phi x$ is not always x, but only has x for *one of its values*. Thus $(\sqrt{x})^2$ is x, but $\sqrt{(x^2)}$ is either x or $-x$, at pleasure, or else one or the other, as dictated by the particular problem in hand. Similarly

$$\cos^{-1}\cos\theta = 2m\pi \pm \theta, \qquad \tan^{-1}\tan\theta = m\pi + \theta,$$

$$\sin^{-1}\sin\theta = 2m\pi + \theta \text{ or } (2m+1)\pi - \theta, \quad \cot^{-1}\cot\theta = m\pi + \theta,$$

$$\cos^{-1}\sin\theta = 2m\pi \pm \left(\frac{\pi}{2} - \theta\right), \qquad \tan^{-1}\cot\theta = m\pi + \frac{\pi}{2} - \theta, \text{ \&c.}$$

This chapter is wholly on expression, and is intended to enable the student to understand the theorems hitherto demonstrated, when expressed in inverse language. All that is wanted, then, is a set of examples for consideration. I shall give two cases with full explanation, and then write down others to be considered by the student.

$$\cos(2\sin^{-1}x) = 1 - 2x^2.$$

This is nothing more than $\cos 2\theta = 1 - 2\sin^2\theta$. Let x be the sine of θ, that is, let θ be an angle whose sine is x, and substitute. The formula is to be understood as 'the cosine of double *any* angle whose sine is x is $1 - 2x^2$.'

$$\tan^{-1}x + \tan^{-1}y = \tan^{-1}\left(\frac{x+y}{1-xy}\right).$$

In all the formulæ which have inverse functions for their terms, we have choice on one side and not on the other. In the formula $4^{\frac{1}{2}} \times 9^{\frac{1}{2}} = 36^{\frac{1}{2}}$ we are not at liberty to say that *any* square root of 4 multiplied by *any* square root of 9 is *any* square root of 36: but *any* square root of 4 multiplied by *any* square root of 9 is *one* of the square roots of 36. And by the above we mean that *any* angle whose tangent is x augmented by *any* angle whose tangent is y, is *one* of the angles whose tangent is $(x + y) \div (1 - xy)$. It is proved thus:

$$\tan(\phi + \theta) = \frac{\tan\phi + \tan\theta}{1 - \tan\phi \, . \tan\theta}, \quad \phi + \theta = \tan^{-1}\left(\frac{\tan\phi + \tan\theta}{1 - \tan\phi \, . \tan\theta}\right),$$

or $\phi + \theta$ is *one* of the angles, &c. Let $\tan\phi = x$, $\tan\theta = y$, or let $\phi = \tan^{-1}x$, $\theta = \tan^{-1}y$, and substitute.

The student may now employ himself on the following:

$$\sin\cos^{-1}x = \sqrt{(1-x^2)},\quad \tan\sec^{-1}x = \sqrt{(x^2-1)},\quad \sin(2\sin^{-1}x) = 2x\sqrt{(1-x^2)},$$

$$\sin(3\sin^{-1}x) = 3x - 4x^3,\quad \sin(4\sin^{-1}x) = (4x - 8x^3)\sqrt{(1-x^2)},$$

$$\tan(2\tan^{-1}x) = \frac{2x}{1-x^2},\quad \tan(3\tan^{-1}x) = \frac{3x - x^3}{1 - 3x^2},$$

taking acute angles,

$$\frac{\pi}{4} = \tan^{-1}\tfrac{1}{2} + \tan^{-1}\tfrac{1}{3} = 4\tan^{-1}\tfrac{1}{5} - \tan^{-1}\tfrac{1}{239}$$

$$= 4\tan^{-1}\tfrac{1}{5} - \tan^{-1}\tfrac{1}{70} + \tan^{-1}\tfrac{1}{99},$$

$$\sin^{-1}x + \sin^{-1}y = \sin^{-1}\{x\sqrt{(1-y^2)} + y\sqrt{(1-x^2)}\},$$

$$\cos^{-1}x + \cos^{-1}y = \cos^{-1}\{\sqrt{(1 - x^2 - y^2 + x^2y^2)} - xy\},$$

$$\cos\tan^{-1}\sin\cot^{-1}x = \sqrt{\frac{x^2+1}{x^2+2}},\quad \cos.\sin^{-1}.\cos.\sin^{-1}x = \pm x,$$

$$\cos.\sec^{-1}.\sin.\tan^{-1}.\cos.\tan^{-1}.\sin\cos^{-1}.\tan.\sin^{-1}x = \sqrt{\frac{3-4x^2}{1-x^2}},$$

$$\sin 2\cos^{-1}\tan 3\cot^{-1}x = \frac{(6x^2 - 2)\sqrt{(x^6 - 15x^4 + 15x^2 - 1)}}{x^2(x^2-3)^2}.$$

This is a chapter on language; and some of the preceding examples are merely hard phrases to be construed from trigonometry into algebra. But such transformations have an important use in calculation. If we wanted to calculate the value of the last-named function of x when $x = 5{\cdot}1761328$, and had such trigonometrical tables as those of Hutton, hereafter described, it would be the easiest plan, beyond comparison, to proceed by the first side. That is, we should find by the table the angle whose cotangent is 5·1761328, treble it, find the tangent of the trebled angle from the table, pass to the table of cosines with that tangent, find the angle to it, double that angle, and take the sine of the last. Thus $\sin\cos^{-1}x$ is, with tables, easier than $\sqrt{(1-x^2)}$, and $\sin 2\sin^{-1}x$ easier than $2x\sqrt{(1-x^2)}$.

The following are a few instances of reduction to mixed trigonometrical forms:

$$\sqrt{(a^2 + b^2 - 2ab\cos C)} = \sqrt{\{(a+b)^2 - 2ab(1+\cos C)\}}$$

$$= (a+b)\sqrt{\left\{1 - \frac{4ab\cos^2\frac{1}{2}C}{(a+b)^2}\right\}} = (a+b)\cos\sin^{-1}\left(\frac{2\sqrt{ab}.\cos\frac{1}{2}C}{a+b}\right),$$

$$\sqrt{(a^2+b^2-2ab\cos C)} = \sqrt{\{(a-b)^2+2ab\,(1-\cos C)\}}$$

$$= (a-b)\sqrt{\left\{1+\frac{4ab\sin^2\frac{1}{2}C}{(a-b)^2}\right\}} = (a-b)\sec\tan^{-1}\left(\frac{2\sqrt{ab}\,.\sin\frac{1}{2}C}{a-b}\right),$$

$$\sqrt{(a^2+b^2)} = a\sec\tan^{-1}\frac{b}{a},\quad \sqrt{(a^2-b^2)} = a\cos\sin^{-1}\frac{b}{a},$$

$$\frac{-b\pm\sqrt{(b^2-4ac)}}{2a} = -\frac{b}{a}\cos^2\tfrac{1}{2}\sin^{-1}\frac{2\sqrt{ac}}{b},\quad \text{or } -\frac{b}{a}\sin^2\tfrac{1}{2}\sin^{-1}\frac{2\sqrt{ac}}{b}.$$

CHAPTER V.

INTRODUCTION OF THE UNEXPLAINED SYMBOL $\sqrt{-1}$.

If we look at the series for $\sin\theta$ and $\cos\theta$, of the form obtained in p. 33, we see that each term is one of those in ϵ^θ (*Algebra*, p. 225). We easily deduce

$$\cos\theta + k\sin\theta = 1 + k\theta - \frac{\theta^2}{2} - k\frac{\theta^3}{2.3} + \frac{\theta^4}{2.3.4} + k\frac{\theta^5}{2.3.4.5} - \dots.$$

If there existed such a quantity k, as would give $k^2 = -1$, $k^3 = -k$, $k^4 = 1$, $k^5 = k$, &c., then $\cos\theta + k\sin\theta$ would be $\epsilon^{k\theta}$. Such a quantity there is not in Algebra, as hitherto considered: for $k^2 = -1$ is absurd. If, under pretence of satisfying this equation, we invent $k = \sqrt{-1}$, and proceed to use it according to the laws which demonstrably govern our intelligible symbols of positive and negative quantity, we adopt the process of all the algebraists, with a fair statement of what we are doing. A use, which ought to have been called *experimental*, of the *symbol* $\sqrt{-1}$, under the name of an *impossible* quantity, shewed that, come how it might, the intelligible results (when such things occurred) of the experiment were always true, and otherwise demonstrable. I am now going to try some of these experiments: the student may rest assured that the *new* results of this chapter will, in the second book, be rendered demonstrative, upon a system which clearly defines $\sqrt{-1}$; or he may doubt it: but he must not think they are demonstrated *here*, though they will have strong moral* evidence in their favour. By giving precedence to the *use* of $\sqrt{-1}$, under the above stipulation, the student will gain the advantage of familiarity with the language of double algebra, before he approaches the difficulties.

* It is almost impossible to discredit Woodhouse's remark;—"Whether I have found a logic, by the rules of which operations with imaginary quantities are conducted, is not now the question: but surely this is evident, that since they lead to right conclusions, they *must have a logic*."

Say that we suppose, from the above,

$$\epsilon^{\theta\sqrt{-1}} = \cos\theta + \sin\theta \,.\, \sqrt{-1}.$$

The processes of algebra constantly lead to this result, and refuse every other; I mean those in which $\sqrt{-1}$ is assumed to be something which, though unintelligible, is governed by the laws of algebra—a fellow-subject of the other symbols, with a mask over his features. For instance, common multiplication will give

$$(\cos\theta + \sin\theta \,.\sqrt{-1})\,(\cos\phi + \sin\phi \,.\sqrt{-1}) = \cos(\phi+\theta) + \sin(\phi+\theta) \,.\sqrt{-1}.$$

Let $f\theta$ denote $\cos\theta + \sin\theta \,.\sqrt{-1}$; then $f\theta \times f\phi = f(\theta+\phi)$, and (*Algebra*, p. 204) $f\theta$ must be E^θ, where E is independent of θ. Accordingly,

$$\frac{E^\theta - 1}{\theta} = \frac{\cos\theta - 1}{\theta} + \frac{\sin\theta}{\theta} \,.\sqrt{-1}.$$

Diminish θ without limit, and (p. 17, and Algebra, p. 266)

$$\log E = 0 + \sqrt{-1}, \quad \text{or} \quad E = \epsilon^{\sqrt{-1}}, \quad E^\theta = \epsilon^{\theta\sqrt{-1}}.$$

If $\epsilon^{\theta\sqrt{-1}} = \cos\theta + \sin\theta \,.\sqrt{-1}$, universally, then

$$\epsilon^{-\theta\sqrt{-1}} = \cos\theta - \sin\theta \,.\sqrt{-1}, \text{ whence}$$

$$\cos\theta = \frac{\epsilon^{\theta\sqrt{-1}} + \epsilon^{-\theta\sqrt{-1}}}{2}, \quad \sin\theta = \frac{\epsilon^{\theta\sqrt{-1}} - \epsilon^{-\theta\sqrt{-1}}}{2\sqrt{-1}}.$$

Had these forms been intelligible, they would have been the proper *algebraical* definitions of the *cosine* and *sine* of θ; and trigonometry would have been pure algebra in the ancient sense, and a very easy part of it. For assuming $\tan\theta$ to be $\sin\theta \div \cos\theta$, and $\sec\theta$, $\operatorname{cosec}\theta$, $\cot\theta$, to be reciprocals of the other three, all the formulæ of trigonometry would have been proved by simple algebraical operation. For example,

$$\begin{aligned}\cos\phi\,\cos\theta &= \tfrac{1}{4}\left(\epsilon^{\phi\sqrt{-1}} + \epsilon^{-\phi\sqrt{-1}}\right)\left(\epsilon^{\theta\sqrt{-1}} + \epsilon^{-\theta\sqrt{-1}}\right)\\ &= \tfrac{1}{4}\left(\epsilon^{(\phi+\theta)\sqrt{-1}} + \epsilon^{-(\phi+\theta)\sqrt{-1}} + \epsilon^{(\phi-\theta)\sqrt{-1}} + \epsilon^{-(\phi-\theta)\sqrt{-1}}\right)\\ &= \tfrac{1}{2}\{\cos(\phi+\theta) + \cos(\phi-\theta)\}.\end{aligned}$$

Since $\epsilon^{\theta\sqrt{-1}}$ has this property, that a change of θ into $n\theta$ raises it to the nth power, we must have

$$(\cos\theta + \sin\theta \,.\sqrt{-1})^n = \cos n\theta + \sin n\theta \,.\sqrt{-1}.$$

This is called *De Moivre's Theorem*. The student, instead of referring to it, must take pains to associate $\cos\theta + \sin\theta \,.\sqrt{-1}$ with the notion of a quantity which is squared, cubed, &c., by introducing a double, treble, &c. angle. And in like manner he must associate the notion of *reciprocals* with $\cos\theta + \sin\theta \,.\sqrt{-1}$

and $\cos\theta - \sin\theta . \sqrt{-1}$, without being obliged to bring them back to $\epsilon^{\theta\sqrt{-1}}$ and $\epsilon^{-\theta\sqrt{-1}}$. The following equation will assist:

$$(\cos\theta + \sin\theta . \sqrt{-1})\ (\cos\theta - \sin\theta . \sqrt{-1}) = \cos^2\theta + \sin^2\theta = 1.$$

If n be integer, the first side has one value only, and also the second. But if n be fractional, as $\frac{7}{10}$, the first side has 10 distinct values; the second apparently only one. This introduces us to a new consideration of the highest importance.

We have been using an angle in two different ways: first, absolutely, as a magnitude, in the same manner as any other kind of magnitude; secondly, as generated by a straight line revolving from a given starting-line, and indicating the direction which the revolving line points out when it has revolved through the angle. As magnitudes, θ and $\theta + 2\pi$ arcual units of angle differ as much as θ and $\theta + 2\pi$ feet, or gallons, or hours: as indicators of direction, they yield no difference at all—they indicate the same direction.

If we begin with θ as indicating a direction (for which $\theta + 2m\pi$ would have done as well), and if $F\theta$ be the solution of a problem in which θ is a given quantity, that problem is equally solved by $F(\theta + 2m\pi)$, m being any integer positive or negative. So many different values as we can give to $F(\theta + 2m\pi)$, so many different solutions: but if $F\theta$ be another angle, used as an indicator of direction, then so many different values as we can find for $F(\theta + 2m\pi)$, no two of which differ by a positive or negative multiple of 2π, so many distinct answers are indicated. And all that we say of directions applies to the trigonometrical functions, which take value only from the direction of the revolving line, and not at all from the number of revolutions by which it has been attained.

If θ be an angle which indicates a direction, $n\theta$ can only indicate *one* direction, when n is *integer*. For, using $\theta + 2m\pi$ for θ, $n\theta$ becomes $n\theta + 2nm\pi$, and $2nm$ is an even integer. But if n be a commensurable arithmetical fraction, say $\frac{p}{q}$ in its lowest terms, then $n\theta$ indicates q distinct directions, no more and no fewer. For $n(\theta + 2m\pi)$ or $\frac{p}{q}\theta + \frac{2mp}{q}\pi$ indicates the same direction for any two values of m, m' and m'', in which $\frac{2m'p}{q} - \frac{2m''p}{q}$

is an even integer, or $(m' - m'')\frac{p}{q}$ an integer, positive or negative; and for no others. Now since q and p are prime to each other, this can only be when q divides $m' - m''$, positively or negatively, or when m' and m'' differ by a multiple of q.

If then we take the following values of m, namely,

$$0, 1, 2, 3, \ldots (q-1),$$

we get all that give really different directions; for every other number, positive or negative, differs from one or another of these by a multiple of q. All the distinct directions, then, are indicated by one or another of the following angles,

$$\frac{p}{q}\theta,\ \frac{p}{q}\theta + \frac{p}{q}2\pi,\ \frac{p}{q}\theta + \frac{2p}{q}2\pi,\ \frac{p}{q}\theta + \frac{3p}{q}2\pi \ldots . \frac{p}{q}\theta + \frac{(q-1)p}{q}2\pi.$$

Now if p and q be prime to one another (Arithmetic, Appendix, p. 195), and if we divide p, $2p$, ... $(q-1)p$ severally by q, the remainders (which are all we need look to, since every unit in a quotient is 2π in the angle), whatever order they may occur in, are 1, 2, 3, ... $(q-1)$, each occurring once somewhere. Consequently, changing the order, we may say that all the directions which $n(\theta + 2m\pi)$ can indicate, are those indicated by

$$\frac{p}{q}\theta,\ \frac{p}{q}\theta + \frac{1}{q}.2\pi,\ \frac{p}{q}\theta + \frac{2}{q}.2\pi,\ \ldots \frac{p}{q}\theta + \frac{q-1}{q}.2\pi,$$

which may be expressed thus: If q be the lowest denominator of n, all the directions indicated by $n\theta$ may be derived from any one of them, by successive advances of the qth part of a revolution each.

In the last equation, written thus,

$$(\cos\theta + \sin\theta.\sqrt{-1})^{\frac{p}{q}} = \cos\frac{p}{q}\theta + \sin\frac{p}{q}\theta.\sqrt{-1},$$

we see q different results on the second side; which are the q ambiguities of value we are taught by common algebra to give to the first side.

If $\cos a + \sin a.\sqrt{-1}$ be one of the values of $(\cos\theta + \sin\theta.\sqrt{-1})^n$, then $\cos a - \sin a.\sqrt{-1}$ is, by similar reasoning, one of the values of $(\cos\theta - \sin\theta.\sqrt{-1})^n$. If $\sin\theta = 0$, that is, if $\cos\theta$ be either $+1$ or -1, then $\cos\theta + \sin\theta.\sqrt{-1}$ is the same as $\cos\theta - \sin\theta.\sqrt{-1}$, or both $\cos a + \sin a . \sqrt{-1}$ and $\cos a - \sin a . \sqrt{-1}$ are values of both. This will enable us to arrange our sets with better perception of their connexion.

I shall now exhibit all the 12th roots of $+1$, and all the 12th roots of -1. For the first, $\theta = 0$, for the second $\theta = \pi$, (which gives $\frac{1}{12}\pi$ for a commencement), and on these angles we must make advances of one-twelfth of a revolution at each step, stopping when we have gained 12 distinct values for each. Gradual measurement will here be most convenient.

The twelve twelfth roots of + 1.

$\cos 0° \pm \sin 0° . \sqrt{-1}$ gives only one root, + 1 itself

$\cos 30° \pm \sin 30° . \sqrt{-1}$ ...... $\frac{1}{2}\sqrt{3} \pm \frac{1}{2}.\sqrt{-1}$

$\cos 60° \pm \sin 60° . \sqrt{-1}$ ...... $\frac{1}{2} \pm \frac{1}{2}\sqrt{3}.\sqrt{-1}$

$\cos 90° \pm \sin 90° . \sqrt{-1}$ ...... $\pm \sqrt{-1}$

$\cos 120° \pm \sin 120° . \sqrt{-1}$ ...... $-\frac{1}{2} \pm \frac{1}{2}\sqrt{3}.\sqrt{-1}$

$\cos 150° \pm \sin 150° . \sqrt{-1}$ ...... $-\frac{1}{2}\sqrt{3} \pm \frac{1}{2}.\sqrt{-1}$

$\cos 180° \pm \sin 180° . \sqrt{-1}$ ...... only one root, -1.

The twelve twelfth roots of − 1.

$\cos 15° \pm \sin 15° . \sqrt{-1}$ gives $\frac{\sqrt{6}+\sqrt{2}}{4} \pm \frac{\sqrt{6}-\sqrt{2}}{4}.\sqrt{-1}$

$\cos 45° \pm \sin 45° . \sqrt{-1}$ ...... $\frac{1}{2}\sqrt{2} \pm \frac{1}{2}\sqrt{2}.\sqrt{-1}$

$\cos 75° \pm \sin 75° . \sqrt{-1}$ ...... $\frac{\sqrt{6}-\sqrt{2}}{4} \pm \frac{\sqrt{6}+\sqrt{2}}{4}.\sqrt{-1}$

$\cos 105° \pm \sin 105° . \sqrt{-1}$ ...... $-\frac{\sqrt{6}-\sqrt{2}}{4} \pm \frac{\sqrt{6}+\sqrt{2}}{4}.\sqrt{-1}$

$\cos 135° \pm \sin 135° . \sqrt{-1}$ ...... $-\frac{1}{2}\sqrt{2} \pm \frac{1}{2}\sqrt{2}.\sqrt{-1}$

$\cos 165° \pm \sin 165° . \sqrt{-1}$ ...... $-\frac{\sqrt{6}+\sqrt{2}}{4} \pm \frac{\sqrt{6}-\sqrt{2}}{4}.\sqrt{-1}$.

We have now found the twelfth roots of any quantity, positive or negative. If a be any twelfth root of $+1$, $a\sqrt[12]{m}$, m being positive, is a twelfth root of m. For its twelfth power is $a^{12}.m$, or m. Similarly, if β be any twelfth root of -1, $\beta\sqrt[12]{m}$ is a twelfth root of $-m$.

We may extend this further, as follows. Two given quantities a, and b, cannot be cosine and sine to the same angle, unless $a^2+b^2=1$: but they may be *proportional* to the sine and cosine of the same angle. For if $a^2+b^2=m^2$, then $\frac{a}{m}$ and $\frac{b}{m}$ are cosine and sine to the same angle, and the tangent of that angle is $\frac{b}{a}$. Hence we have a transformation of great importance,

$$a+b\sqrt{-1}=\sqrt{(a^2+b^2)}.\left(\cos\tan^{-1}\frac{b}{a}+\sin\tan^{-1}\frac{b}{a}.\sqrt{-1}\right)=\sqrt{(a^2+b^2)}\epsilon^{\tan^{-1}\frac{b}{a}\sqrt{-1}}.$$

But this point is to be remembered: $\tan^{-1}x$ has two values which indicate different directions; and those values are opponents; θ being one, $\theta+\pi$ is the other. Now θ and $\theta+\pi$ have contrary sines and cosines; $\sin(\theta+\pi)=-\sin\theta$, $\cos(\theta+\pi)=-\cos\theta$. If we set out with a given sign, say the positive one, for $\sqrt{(a^2+b^2)}$, and take the wrong angle, we shall end with $-a-b\sqrt{-1}$, instead of $+a+b\sqrt{-1}$: we may set it right either by altering the angle, or using the other square root of a^2+b^2. But as the positive root is generally used, the proper value of $\tan^{-1}\frac{b}{a}$ may be remembered as the angle whose cosine has the same sign as a. The following would be the most convenient arrangement. One angle which has $\frac{b}{a}$ for tangent has its sine of the same sign as a, and its cosine of the same sign as b; the other has the sine of a different sign from a, and the cosine of a different sign from b. Let the first be denoted by $\tan^{-1}\frac{b}{a}$, and the second by $\tan^{-1}\frac{-b}{-a}$. Thus we have

$$a+b\sqrt{-1}=\sqrt{(a^2+b^2)}.\epsilon^{\tan^{-1}\frac{b}{a}.\sqrt{-1}}=-\sqrt{(a^2+b^2)}.\epsilon^{\tan\frac{-b}{-a}.\sqrt{-1}}.$$

Show now that one twelfth root of $a+b\sqrt{-1}$ is

$$\sqrt[12]{\sqrt{(a^2+b^2)}}.\left\{\cos\left(\tfrac{1}{12}\tan^{-1}\frac{b}{a}\right)+\sin\left(\tfrac{1}{12}\tan^{-1}\frac{b}{a}\right).\sqrt{-1}\right\};$$

and that all the twelfth roots may be found, either by successively increasing the angle by twelfths of a revolution, or by multiplying the above by all the twelfth roots of $+1$.

Returning to our original notation, we have

$$x + y\sqrt{-1} = r(\cos\theta + \sin\theta.\sqrt{-1}) = r\epsilon^{\theta\sqrt{-1}};$$

$$x - y\sqrt{-1} = r(\cos\theta - \sin\theta.\sqrt{-1}) = r\epsilon^{-\theta\sqrt{-1}}.$$

The fundamental equation $\epsilon^{\theta\sqrt{-1}} = \cos\theta + \sin\theta.\sqrt{-1}$, gives the following results; $\epsilon^{2m\pi.\sqrt{-1}} = 1$, and $\epsilon^{(2m+1)\pi.\sqrt{-1}} = -1$. If a be a positive quantity, we have $a = \epsilon^{\log a} = \epsilon^{\log a + 2m\pi.\sqrt{-1}}$. If then any symbol z be a logarithm of a, which satisfies $a = \epsilon^z$, we have such right as we can take in *this* chapter to say that the ordinary arithmetical logarithm (which we shall still denote by $\log a$) is only one of a class, all contained in $\log a + 2m\pi\sqrt{-1}$, in which m may be any positive or negative integer. Let λa denote any logarithm of a; then we have $\lambda a = \log a + 2m\pi\sqrt{-1}$: and the usual form $\log a$ is one case of λa. Here a was positive.

Now $-a = (-1).a = \epsilon^{(2m+1)\pi\sqrt{-1}}.\epsilon^{\log a} = \epsilon^{\log a + (2m+1)\pi\sqrt{-1}}$; whence we may say that negative quantities have logarithms, and that

$$\lambda(-a) = \log a + (2m+1)\pi\sqrt{-1};$$

but still, as before, there is no *arithmetical* logarithm to a negative quantity; for, m being integer, $2m + 1$ cannot vanish.

Let the student now show that, in this extension, *any* logarithm of a, added to *any* logarithm of b, gives *one* of the logarithms of ab, &c. All our ordinary logarithmic relations remain true in this sense. Since

$$x + y\sqrt{-1} = r\epsilon^{\theta\sqrt{-1}} = r\epsilon^{(\theta+2m\pi).\sqrt{-1}} = \epsilon^{\log r + (\theta+2m\pi).\sqrt{-1}},$$

we have all the system of logarithms exhibited in

$$\lambda(x + y\sqrt{-1}) = \log r + (\theta + 2m\pi)\sqrt{-1},$$

$$= \tfrac{1}{2}\log(x^2 + y^2) + \left(\tan^{-1}\frac{y}{x} + 2m\pi\right)\sqrt{-1}.$$

Here $2m\pi$ is not necessary, unless we restrict $\tan^{-1}\frac{y}{x}$ to be in the first revolution; otherwise, $\tan^{-1}\frac{y}{x}$ $\left(\text{remembering the distinction between it and } \tan^{-1}\frac{-y}{-x}\right)$ expresses every case by itself.

A still further extension of the notion of a logarithm may now be made. The base ϵ is $\epsilon^{1+2m\pi\sqrt{-1}}$. If La represent the most extensive logarithm of a, which we can get from any such form of ϵ, it must be obtained (a being positive) thus:

$$\left(\epsilon^{1+2m\pi\sqrt{-1}}\right)^{La} = \epsilon^{\log a+2n\pi\sqrt{-1}}, \quad \left(\epsilon^{1+2m\pi}\right)^{L(-a)} = \epsilon^{\log a+(2n+1)\pi\sqrt{-1}};$$

or $$La = \frac{\log a + 2n\pi\sqrt{-1}}{1+2m\pi\sqrt{-1}}, \quad L(-a) = \frac{\log a + (2n+1)\pi\sqrt{-1}}{1+2m\pi\sqrt{-1}};$$

m and n being any positive or negative integers. Proceed in the same way with $x + y\sqrt{-1}$, and we have

$$L(x+y\sqrt{-1}) = \frac{\log r + (\theta + 2n\pi)\sqrt{-1}}{1+2m\pi\sqrt{-1}}.$$

It thus appears that there is, to the base ϵ, an infinite number of systems of logarithms, corresponding to the values of m, and an infinite number of logarithms in each system, corresponding to the value of n.

Two logarithms of one quantity, taken out of different *systems*, cannot generally be found equal. If, m and m' being two integers, we form the equation {let $p = \theta + 2\pi n$, $p' = \theta + 2\pi n'$}

$$\frac{\log r + p\sqrt{-1}}{1+2m\pi\sqrt{-1}} = \frac{\log r + p'\sqrt{-1}}{1+2m'\pi\sqrt{-1}},$$

clear it of fractions, and equate the possible and impossible terms, we get $pm' = p'm$ and $2m'\pi \log r + p = 2m\pi \log r + p'$. Substitute in the second the value of p' from the first, and we get

$$(2\pi m \log r - p)(m' - m) = 0;$$

either then $m' = m$, and the systems are the same, or $2\pi m \log r = p$ and $2\pi m' \log r = p'$. In these cases a logarithm of $x + y\sqrt{-1}$ or $r\epsilon^{\theta\sqrt{-1}}$, in each system is $\log r$, the *arithmetical* logarithm of r. But, m and m' being the indices of the bases, and n and n' those of the particular logarithms to those bases, this requires that r and θ should be determined by

$$\frac{\theta+2\pi n}{2\pi m} = \frac{\theta + 2\pi n'}{2\pi m'}, \quad \text{and } \log r = \frac{\theta + 2\pi n}{2\pi m},$$

$$\text{or } \log r = \frac{n'-n}{m'-m}, \quad \theta = 2\pi\frac{n'm - nm'}{m'-m};$$

so that, for two given systems, and two given values of n in those systems, there is one expression, and one only, which has the same logarithms in both.

In ordinary algebra it is said that *negative quantities have none but impossible logarithms.* And this in the face of the result that, $\sqrt{\varepsilon}$ being a, $\varepsilon^{\frac{1}{2}}$ is either $+a$ or $-a$, so that $-a$ has $\frac{1}{2}$ for a logarithm. We can now show how these isolated cases of negative quantities with real logarithms arise.

Let us solve the general question—What expressions have real logarithms, what are they, and in what systems? The following equation is produced by multiplying both terms of the fraction by $1-2m\pi\sqrt{-1}$,

$$L(x+y\sqrt{-1})=\frac{\log r+(\theta+2n\pi)\sqrt{-1}}{1+2m\pi\sqrt{-1}}$$

$$=\frac{\log r+2m\pi(\theta+2n\pi)}{1+4m^2\pi^2}+\frac{\theta+2n\pi-2m\pi\log r}{1+4m^2\pi^2}\sqrt{-1}.$$

This is a real quantity only when $2m\pi\log r=\theta+2n\pi$, in which case $L(x+y\sqrt{-1})=\log r(1+4m^2\pi^2)\div(1+4m^2\pi^2)=\log r$. If $\theta=\pi$, in which case $y=0$ and x is negative $(x=r\epsilon^{\pi\sqrt{-1}}=-r)$ we have $\log(-r)=\log r$, whenever $2m\pi\log r=(2n+1)\pi$ or $r=\varepsilon^{\frac{2n+1}{2m}}$. This is precisely the case we might have anticipated: for ε^{2n+1} has two real $(2m)^{\text{th}}$ roots, one negative. But it appears that instead of the system being that of the base ε, the base is $\varepsilon^{1+2m\pi\sqrt{-1}}$. The complete illustration of this difficulty may be gathered from the second book.

Returning now to the fundamental equations, let z stand for $\epsilon^{\theta\sqrt{-1}}$ or $\cos\theta+\sin\theta.\sqrt{-1}$. We have then

$$z=\cos\theta+\sin\theta.\sqrt{-1},\qquad z^n=\cos n\theta+\sin n\theta.\sqrt{-1},$$

$$z^{-1}=\cos\theta-\sin\theta.\sqrt{-1},\qquad z^{-n}=\cos n\theta-\sin n\theta.\sqrt{-1},$$

$$2\cos\theta=z+z^{-1},\qquad 2\cos n\theta=z^n+z^{-n},$$

$$2\sqrt{-1}.\sin\theta=z-z^{-1},\qquad 2\sqrt{-1}.\sin n\theta=z^n-z^{-n}.$$

These equations may almost be said to contain trigonometry. Completely established, they would furnish proof of all we have done: the deduction from them of our previous results must be inductive proof that, somehow or other, the use of $\sqrt{-1}$ does lead to true results.

Required $\sin^3\theta$ in terms of functions of multiples of θ:

$$\sin^3\theta=\left(\frac{1}{2\sqrt{-1}}\right)^3\{z-z^{-1}\}^3=-\frac{1}{8\sqrt{-1}}\{z^3-3z+3z^{-1}-z^{-3}\}$$

$$=-\tfrac{1}{4}\{\sin 3\theta-3\sin\theta\},\text{ as in page 35.}$$

Required $\cos 4\theta$ in powers of $\sin\theta$ and $\cos\theta$:

$$\cos 4\theta = \frac{z^4 + z^{-4}}{2} = \frac{(c + s\sqrt{-1})^4 + (c - s\sqrt{-1})^4}{2}$$

$$= c^4 - 6c^2s^2 + s^4, \text{ as in page 31.}$$

The student must take notice of the manner in which the *action* of $\sqrt{-1}$ supplies the place of the rule in page 31.

We may now extend some of our rules. Required $\sin^m\theta \cos^n\theta$ in functions of multiples of θ. That is, we have to find

$$\frac{(z - z^{-1})^m (z + z^{-1})^n}{2^{m+n} (\sqrt{-1})^m}.$$

In the numerator, the descent of each developed factor is by two dimensions in each term; for

$$(z - z^{-1})^m \text{ is } z^m - 1_m z^{m-1}z^{-1} + 2_m z^{m-2}z^{-2} - \ldots\ldots$$

Now if we multiply $Ka^k + La^{k-1}b + Ma^{k-2}b^2 + \ldots + Pab^{k-1} + Qb^k$ by $a - b$ or by $a + b$, we have as results,

$$(K-0)\,a^{k+1} + (L-K)a^kb + (M-L)\,a^{k-1}b^2 + \ldots + (Q-P)\,ab^k + (0-Q)b^{k+1},$$

$$(K+0)\,a^{k+1} + (L+K)a^kb + (M+L)a^{k-1}b^2 + \ldots + (Q+P)\,ab^k + (0+Q)\,b^{k+1}.$$

The shortest way of doing this is by writing down the coefficients K, L, &c. in a row, and under them $K \mp 0$, $L \mp K$, &c. In this manner we may rapidly make the multiplications, and in either of two mutually verificatory ways: by coefficients from $(z - z^{-1})^m$, and successive multiplications by $z + z^{-1}$, or by coefficients from $(z + z^{-1})^n$ and successive multiplications by $z - z^{-1}$. As to the rest, the final divisor will be 2^{m+n-1}, for a 2 (and $\sqrt{-1}$, if there) will be taken up in the reconversion of $z^m \pm z^{-m}$ into cosine or sine. And if m be even-even, $(\sqrt{-1})^m$ is 1, and the result is in cosines; if odd-odd, it is $\sqrt{-1}$, and the result is in sines; if odd-even, it is -1, and the result is, we may say, in negative cosines, the sign of each term being changed; if even-odd, it is $-\sqrt{-1}$, and the result is in negative sines. And the conclusion begins with $(m + n)\,\theta$ and this angle diminishes by 2θ at each step. But if z^0 occur, there is no term distinct having z^{-0}, so that the 2 just mentioned is not taken up in forming $\cos 0\theta$, and must therefore be used in denoting the coefficient; or only half the coefficient in the result must be used. As an instance, I shall take $\sin^5\theta \cos^6\theta$, in which, if we work both ways for verification, we shall pick up during the process all that is wanting for finding $\sin^m\theta \cos^6\theta$, for any value of m not exceeding 5, and $\sin^5\theta\cos^n\theta$, for any value of n not exceeding 6. The conclusion which each step prepares us for is written in abbreviation at the commencement.

c^6	$1+6+15+20+15+\ 6+\ 1$
$s\,c^6$	$1+5+\ 9+\ 5-\ 5-\ 9-\ 5-\ 1$
s^2c^6	$1+4+\ 4-\ 4-10-\ 4+\ 4+\ 4+1$
s^3c^6	$1+3+\ 0-\ 8-\ 6+\ 6+\ 8+\ 0-3-1$
s^4c^6	$1+2-\ 3-\ 8+\ 2+12+\ 2-\ 8-3+2+1$
s^5c^6	$1+1-\ 5-\ 5+10+10-10-10+5+5-1-1$

s^5	$1-5+10-10+\ 5-\ 1$
s^5c	$1-4+\ 5+\ 0-\ 5+\ 4-\ 1$
s^5c^2	$1-3+\ 1+\ 5-\ 5-\ 1+\ 3-\ 1$
s^5c^3	$1-2-\ 2+\ 6+\ 0-\ 6+\ 2+\ 2-1$
s^5c^4	$1-1-\ 4+\ 4+\ 6-\ 6-\ 4+\ 4+1-1$
s^5c^5	$1+0-\ 5+\ 0+10+\ 0-10+\ 0+5+0-1$
s^5c^6	$1+1-\ 5-\ 5+10+10-10-10+5+5-1-1$

Attending to the rest of the process, we have

$$2^5\cos^6\theta = \cos 6\theta + 6\cos 4\theta + 15\cos 2\theta + 10$$
$$2^6\sin\theta\cos^6\theta = \sin 7\theta + 5\sin 5\theta + 9\sin 3\theta + 5\sin\theta$$
$$2^7\sin^2\theta\cos^6\theta = -\cos 8\theta - 4\cos 6\theta - 4\cos 4\theta + 4\cos 2\theta + 5$$
$$2^8\sin^3\theta\cos^6\theta = -\sin 9\theta - 3\sin 7\theta + 8\sin 3\theta + 6\sin\theta$$
$$2^9\sin^4\theta\cos^6\theta = \cos 10\theta + 2\cos 8\theta - 3\cos 6\theta - 8\cos 4\theta + 2\cos 2\theta + 6$$
$$2^{10}\sin^5\theta\cos^6\theta = \sin 11\theta + \sin 9\theta - 5\sin 7\theta - 5\sin 5\theta + 10\sin 3\theta + 10\sin\theta$$

$$2^4\sin^5\theta = \sin 5\theta - 5\sin 3\theta + 10\sin\theta$$
$$2^5\sin^5\theta\cos\theta = \sin 6\theta - 4\sin 4\theta + 5\sin 2\theta$$
$$2^6\sin^5\theta\cos^2\theta = \sin 7\theta - 3\sin 5\theta + \sin 3\theta + 5\sin\theta$$
$$2^7\sin^5\theta\cos^3\theta = \sin 8\theta - 2\sin 6\theta - 2\sin 4\theta + 6\sin 2\theta$$
$$2^8\sin^5\theta\cos^4\theta = \sin 9\theta - \sin 7\theta - 4\sin 5\theta + 4\sin 3\theta + 6\sin\theta$$
$$2^9\sin^5\theta\cos^5\theta = \sin 10\theta - 5\sin 6\theta + 10\sin 2\theta$$
$$2^{10}\sin^5\theta\cos^6\theta = \sin 11\theta + \sin 9\theta - 5\sin 7\theta - 5\sin 5\theta + 10\sin 3\theta + 10\sin\theta$$

Let $a_0 + a_1x + a_2x^2 + \ldots$ be a converging series, if extending *ad infinitum* (which it need not here do), and let it be the development of a known function of x, ϕx. It is required to find

$$a_0 + a_1x\cos\theta + a_2x^2\cos 2\theta + \ldots\ldots$$

Multiplied by 2, with $z^n + z^{-n}$ written for $2\cos n\theta$ in the several cases, it obviously becomes $\phi(xz) + \phi(xz^{-1})$. It is then

$$\tfrac{1}{2}\{\phi(xz) + \phi(xz^{-1})\},$$

which can be reduced to a real form. Similarly

$$a_0 \sin 0\theta + a_1 x \sin\theta + a_2 x^2 \sin 2\theta + \dots \text{ is } \{\phi(xz) - \phi(xz^{-1})\} \div 2\sqrt{-1}.$$

Before proceeding to some examples, it will be worth while (seeing before us a field of such extent as the applying the summation of *any* algebraical series to the summation of one in which the terms are severally multiplied by the sine or cosine of the multiple of an angle), to consider the transformation of $\phi(xz)$ and $\phi(xz^{-1})$.

Let there be a function of $\sqrt{a}$ which, if $\sqrt{a}$ had only one value, would itself have only one value. This restriction of value may be, if we please, conventional; for instance, $\sin^{-1}b \,.\, \sqrt{a}$ is such a function, if we *suppose* ourselves restricted to one value of $\sin^{-1}b$. If, then, $F(\sqrt{a})$ can be thrown into the form $P + Q \,.\, \sqrt{a}$, where P and Q are wholly unaffected by the change of $\sqrt{a}$ into $-\sqrt{a}$, $F(-\sqrt{a})$ must be $P - Q \,.\, \sqrt{a}$.

Now it is a proposition to be carefully remembered, that any function of $x + y\sqrt{-1}$, $x' + y'\sqrt{-1}$, &c., can always be reduced to the form $P + Q\sqrt{-1}$, in which P and Q are wholly independent of $\sqrt{-1}$, or are real quantities. In the second book I shall show this independently of all particular cases: at present we must be content with induction. The proposition is clear enough of sums, differences, and products, however varied; and also when division enters, if we look at its reduction to multiplication by

$$\frac{1}{x + y\sqrt{-1}} = \frac{x - y\sqrt{-1}}{x^2 + y^2} = \frac{x}{x^2 + y^2} - \frac{y}{x^2 + y^2} \,.\, \sqrt{-1}.$$

As to powers, we can thus reduce the form $(x + y\sqrt{-1})^{p+q\sqrt{-1}}$; for in this we see $(r\epsilon^{\theta\sqrt{-1}})^{p+q\sqrt{-1}}$, or

$$\epsilon^{(\log r + \theta\sqrt{-1})\,(p+q\sqrt{-1})}, \text{ or } \epsilon^{p\log r - q\theta + (q\log r + p\theta)\sqrt{-1}}, \text{ or}$$

$$\epsilon^{p\log r - q\theta}\cos(q\log r + p\theta) + \epsilon^{p\log r - q\theta}\sin(q\log r + p\theta) \,.\, \sqrt{-1};$$

for $\epsilon^{x+y\sqrt{-1}}$ we have $\epsilon^x\cos y + \epsilon^x \sin y \,.\, \sqrt{-1}$; for $\log(x + y\sqrt{-1})$ we have

$$\log(x+y\sqrt{-1}) = \log\{\sqrt{(x^2+y^2)}\,\epsilon^{\tan^{-1}\frac{y}{x}\,.\,\sqrt{-1}}\} = \tfrac{1}{2}\log(x^2+y^2) + \tan^{-1}\frac{y}{x} \,.\, \sqrt{-1}.$$

If we extend our notion of cosine and sine, taking the exponential forms in p. 42 as definitions, we have

$$\sin(x+y\sqrt{-1}) = \frac{1}{2\sqrt{-1}}\{\epsilon^{x\sqrt{-1}-y} - \epsilon^{-x\sqrt{-1}+y}\}$$
$$= -\tfrac{1}{2}\sqrt{-1}\{\epsilon^{-y}(\cos x + \sin x.\sqrt{-1}) - \epsilon^{y}(\cos x - \sin x.\sqrt{-1})\}$$
$$= \frac{\epsilon^{y}+\epsilon^{-y}}{2}\sin x + \frac{\epsilon^{y}-\epsilon^{-y}}{2}\cos x.\sqrt{-1}$$

We may reduce $\cos(x+y\sqrt{-1})$ and $\tan(x+y\sqrt{-1})$ in like manner. If, with a like extension, we take $\sin^{-1}(x+y\sqrt{-1})$, we may make the reductions as follows:

$$-\theta\sqrt{-1} = \log(\cos\theta - \sin\theta.\sqrt{-1}), \quad \theta = \sqrt{-1}.\log(\cos\theta - \sin\theta.\sqrt{-1}),$$
$$\sin^{-1} x = \sqrt{-1}.\log\{\sqrt{(1-x^2)} - x\sqrt{-1}\},$$
$$\sin^{-1}(x+y\sqrt{-1}) = \sqrt{-1}\log\{\sqrt{(1-x^2+y^2-2xy\sqrt{-1})} - x\sqrt{-1}+y\},$$

the second side of which, by preceding processes, can be reduced as required.

If, in every case, $\phi(x+y\sqrt{-1})$ can be reduced to $P+Q\sqrt{-1}$, in which P and Q are real, then, by our first remark, $\phi(x-y\sqrt{-1})$ can be reduced to $P-Q\sqrt{-1}$, whence

$\tfrac{1}{2}\{\phi(x+y\sqrt{-1})+\phi(x-y\sqrt{-1})\}$ and $\tfrac{1}{2}\sqrt{-1}\{\phi(x+y\sqrt{-1})-\phi(x-y\sqrt{-1})\}$ are real, being P and $-Q$.

We have now to consider $\phi(xz)$ and $\phi(xz^{-1})$, or

$$\phi(x\cos\theta + x\sin\theta.\sqrt{-1}) \text{ and } \phi(x\cos\theta - x\sin\theta.\sqrt{-1}).$$

A few principal examples will here be sufficient.

Let $\phi x = (1+x)^n$,

$$1+x\cos\theta \pm x\sin\theta.\sqrt{-1} = \sqrt{\{(1+x\cos\theta)^2 + x^2\sin^2\theta\}}\,\epsilon^{\pm\tan^{-1}\frac{x\sin\theta}{1+x\cos\theta}.\sqrt{-1}},$$

the nth power of which is

$$(1+2x\cos\theta+x^2)^{\frac{1}{2}n}\left\{\cos n\tan^{-1}\frac{x\sin\theta}{1+x\cos\theta} \pm \sin n\tan^{-1}\frac{x\sin\theta}{1+x\cos\theta}.\sqrt{-1}\right\},$$

$$\frac{\phi(xz)+\phi(xz^{-1})}{2} = (1+2x\cos\theta+x^2)^{\frac{1}{2}n}\cos n\tan^{-1}\frac{x\sin\theta}{1+x\cos\theta},$$

$$\frac{\phi(xz)-\phi(xz^{-1})}{2\sqrt{-1}} = (1+2x\cos\theta+x^2)^{\frac{1}{2}n}\sin n\tan^{-1}\frac{x\sin\theta}{1+x\cos\theta}:$$

and these are the expressions for

$$1+nx\cos\theta + n\frac{n-1}{2}x^2\cos 2\theta + \ldots, \quad nx\sin\theta + n\frac{n-1}{2}x^2\sin 2\theta + \ldots$$

The verification of such results will be useful practice. For instance, it is asserted above that

$$(1+2x\cos\theta+x^2)\cos 2\tan^{-1}\frac{x\sin\theta}{1+x\cos\theta}=1+2x\cos\theta+x^2\cos 2\theta.$$

Now $\cos 2\tan^{-1}a=\cos^2\tan^{-1}a-\sin^2\tan^{-1}a=(1-\tan^2 a)\div(1+\tan^2 a)$.

The first side of the above is then

$$(1+2x\cos\theta+x^2)\frac{(1+x\cos\theta)^2-x^2\sin^2\theta}{(1+x\cos\theta)^2+x^2\sin^2\theta},\ \text{or } 1+2x\cos\theta+(\cos^2\theta-\sin^2\theta)x^2.$$

Let $n=-1$, and change x into $-x$. Show that $\cos(-\tan^{-1}-a)$ is $1:\sqrt{(1+a^2)}$, and that $\sin\{-\tan^{-1}(-a)\}$ is $a\div\sqrt{(1+a^2)}$, and then show that the above expressions give

$$\frac{1-x\cos\theta}{1-2x\cos\theta+x^2}=1+x\cos\theta+x^2\cos 2\theta+x^3\cos 3\theta+\ldots,$$

$$\frac{x\sin\theta}{1-2x\cos\theta+x^2}=\quad x\sin\theta+x^2\sin 2\theta+x^3\cos 3\theta+\ldots.$$

Verify these by the whole method, ϕx being $1\div(1-x)$. Also show the following,

$$\epsilon^{x\cos\theta}\cos(x\sin\theta)=1+x\cos\theta+\frac{x^2\cos 2\theta}{2}+\frac{x^3\cos 3\theta}{2.3}+\ldots,$$

$$\epsilon^{x\cos\theta}\sin(x\sin\theta)=\quad x\sin\theta+\frac{x^2\sin 2\theta}{2}+\frac{x^3\sin 3\theta}{2.3}+\ldots.$$

Let $\phi x=\log(1+x)$. Then

$$\log(1+x\cos\theta+x\sin\theta\,.\sqrt{-1})$$
$$=\frac{1}{2}\log(1+2x\cos\theta+x^2)+\tan^{-1}\frac{x\sin\theta}{1+x\cos\theta}\,.\sqrt{-1},$$

$$\tfrac{1}{2}\log(1+2x\cos\theta+x^2)=x\cos\theta-\frac{x^2}{2}\cos 2\theta+\frac{x^3}{3}\cos 3\theta-\frac{x^4}{4}\cos 4\theta+\ldots,$$

$$\tan^{-1}\frac{x\sin\theta}{1+x\cos\theta}=x\sin\theta-\frac{x^2}{2}\sin 2\theta+\frac{x^3}{3}\sin 3\theta-\frac{x^4}{4}\sin 4\theta+\ldots.$$

If x be >-1 and $<+1$, both these series are convergent, and there is no ambiguity in the first: but there is in the second. The second series, when convergent, has one definite value: which is it of all the values which the first side may bear? It must be the angle which lies between $-\frac{1}{2}\pi$ and $+\frac{1}{2}\pi$: for when x passes from negative to positive through 0, the series does the same.

When x is greater than unity, these series become divergent, and the student should avoid *founding results* upon divergent series, as the question of their legitimacy is disputed upon grounds to which no answer commanding anything like general assent has yet been given. But they may be used as means of discovery, provided that their results be verified by other means before they are considered as established.

If $x = 1$, we have

$$\log\left(2\cos\frac{\theta}{2}\right) = \cos\theta - \frac{\cos 2\theta}{2} + \frac{\cos 3\theta}{3} - \frac{\cos 4\theta}{4} + \ldots,$$

$$\frac{\theta}{2} = \sin\theta - \frac{\sin 2\theta}{2} + \frac{\sin 3\theta}{3} - \frac{\sin 4\theta}{4} + \ldots,$$

and θ must, in the second, lie between $-\pi$ and $+\pi$. These series belong to a peculiar class; they are convergent, but their convergency is not easily established. Their extreme cases often present some algebraical peculiarity. If we divide both sides of the second by θ and diminish θ without limit, we have $\frac{1}{2} = 1 - 1 + 1 - 1 - \ldots$ (Algebra, p. 197). This is not the place, nor even the work, in which to discuss the peculiar character of these series.

Let $\theta = \frac{1}{2}\pi$. The first of the equations becomes

$$\tfrac{1}{2}\log(1 + x^2) = \tfrac{1}{2}x^2 - \tfrac{1}{4}x^4 + \ldots,$$

as well known. But the second becomes

$$\tan^{-1} x = x - \frac{x^3}{3} + \frac{x^5}{5} - \frac{x^7}{7} + \ldots,$$

a remarkable series, both for its simplicity, and for the use to which it has been put. It is convergent when $x > -1$, and not $> +1$; and thus may be made effective when $\tan^{-1}x > -\frac{1}{4}\pi$, and not $> \frac{1}{4}\pi$. When $x = 1$, $\tan^{-1}x = \frac{1}{4}\pi$, we have

$$\frac{\pi}{4} = 1 - \frac{1}{3} + \frac{1}{5} - \frac{1}{7} + \ldots\ldots,$$

being the first calculable form in which π has been directly presented. But this series, though convergent, is very slowly so, (Algebra, page 184) and would require that we should calculate 500 terms before we could be sure of three decimal places. The following is more convergent, derived from $\frac{1}{3}\sqrt{3} = \tan(\frac{1}{6}\pi)$,

$$\frac{\pi}{6} = \tfrac{1}{3}\sqrt{3}\left\{1 - \frac{1}{3}\cdot\frac{1}{3} + \frac{1}{5}\cdot\frac{1}{3^2} - \frac{1}{7}\cdot\frac{1}{3^3} + \ldots\ldots\right\}.$$

But it is best to resolve π or some known fraction of it, into two or more angles whose tangents are known. Thus,

$$\tan^{-1}\tfrac{1}{2} + \tan^{-1}\tfrac{1}{3} = \tfrac{1}{4}\pi,$$

(page 39), gives

$$\frac{\pi}{4} = \frac{1}{2} + \frac{1}{3} - \frac{1}{3}\left(\frac{1}{2^3} + \frac{1}{3^3}\right) + \frac{1}{5}\left(\frac{1}{2^5} + \frac{1}{3^5}\right) - \ldots\ldots,$$

which may be easily calculated, as follows. Write p and q for $\frac{1}{2}$ and $\frac{1}{3}$ (divide by 4 and 9 at every step):

p = ·50000000000	q = ·33333333333
p^3 = ·12500000000	q^3 = ·03703703704
p^5 = ·03125000000	q^5 = ·00411522634
p^7 = ·00781250000	q^7 = ·00045724737
p^9 = ·00195312500	q^9 = ·00005080526
p^{11} = ·00048828125	q^{11} = ·00000564503
p^{13} = ·00012207031	q^{13} = ·00000062723
p^{15} = ·00003051758	q^{15} = ·00000006969
p^{17} = ·00000762940	q^{17} = ·00000000774
p^{19} = ·00000190735	q^{19} = ·00000000086
p^{21} = ·00000047684	q^{21} = ·00000000009
p^{23} = ·00000011921	
p^{25} = ·00000002980	
p^{27} = ·00000000745	
p^{29} = ·00000000186	
p^{31} = ·00000000047	
p^{33} = ·00000000012	
p^{35} = ·00000000003	

Now let $(p^n + q^n) \div n$ be denoted by r_n.

r_1 = ·83333333333	r_3 = ·05401234568
r_5 = ·00707304527	r_7 = ·00118139248
r_9 = ·00022265892	r_{11} = ·00004490239
r_{13} = ·00000943827	r_{15} = ·00000203915
r_{17} = ·00000044924	r_{19} = ·00000010043
r_{21} = ·00000002271	r_{23} = ·00000000518
r_{25} = ·00000000119	r_{27} = ·00000000028
r_{29} = ·00000000006	r_{31} = ·00000000002
·84063894899	·05524078561
	·84063894899
	$\frac{1}{4}\pi$ = ·78539816338
	4
	π = 3·14159265352

which is correct, with the exception of the last place.

The series for $\tan^{-1}x$ may be easily deduced from one among a number of forms which may best be considered together: as follows.

We have seen a remarkable connexion between exponential* forms on the one hand, and trigonometrical forms on the other. Every trigonometrical function has an imaginary exponential one for its equivalent, and every exponential function an imaginary trigonometrical one. Many imaginary forms of one kind are real ones of the other; and the following is such recapitulation and addition as will put all the most useful transformations together.

$$\sin\theta = \frac{\epsilon^{\theta\sqrt{-1}} - \epsilon^{-\theta\sqrt{-1}}}{2\sqrt{-1}}, \quad \frac{\epsilon^{x} - \epsilon^{-x}}{2} = \frac{\sin(x\sqrt{-1})}{\sqrt{-1}},$$

$$\cos\theta = \frac{\epsilon^{\theta\sqrt{-1}} + \epsilon^{-\theta\sqrt{-1}}}{2}, \quad \frac{\epsilon^{x} + \epsilon^{-x}}{2} = \cos(x\sqrt{-1}),$$

$$\tan\theta = \frac{1}{\sqrt{-1}}\,\frac{1 - \epsilon^{-2\theta\sqrt{-1}}}{1 + \epsilon^{-2\theta\sqrt{-1}}} = \frac{1}{\sqrt{-1}}\,\frac{\epsilon^{2\theta\sqrt{-1}} - 1}{\epsilon^{2\theta\sqrt{-1}} + 1}, \quad \frac{\epsilon^{x} - 1}{\epsilon^{x} + 1} = \frac{\tan(\frac{1}{2}x\sqrt{-1})}{\sqrt{-1}}.$$

$$\theta = \log(\cos\theta - \sin\theta.\sqrt{-1}).\sqrt{-1} = -\log(\cos\theta + \sin\theta.\sqrt{-1}).\sqrt{-1}$$

$$\sin^{-1}x = \log\{\sqrt{(1-x^2)} - x\sqrt{-1}\}.\sqrt{-1} = -\log\{\sqrt{(1-x^2)} + x\sqrt{-1}).\sqrt{-1}$$

$$\cos^{-1}x = \log\{x - \sqrt{(x^2-1)}\}.\sqrt{-1} = -\log\{x + \sqrt{(x^2-1)}\}.\sqrt{-1}$$

$$\log\{x + \sqrt{(x^2+1)}\} = \frac{\sin^{-1}(x\sqrt{-1})}{\sqrt{-1}}, \quad \log\{x + \sqrt{(x^2-1)}\} = \cos^{-1}x.\sqrt{-1}$$

$$\epsilon^{2\theta\sqrt{-1}} = \frac{\cos\theta + \sin\theta.\sqrt{-1}}{\cos\theta - \sin\theta.\sqrt{-1}} = \frac{1 + \tan\theta.\sqrt{-1}}{1 - \tan\theta.\sqrt{-1}}$$

$$\tan^{-1}x = \frac{1}{2\sqrt{-1}}\log\frac{1 + x\sqrt{-1}}{1 - x\sqrt{-1}}, \quad \log\frac{1-x}{1+x} = 2\sqrt{-1}.\tan^{-1}(x\sqrt{-1}),$$

$$\epsilon^{x} = \cos(x\sqrt{-1}) - \sin(x\sqrt{-1}).\sqrt{-1}, \quad \log x = 2\sqrt{-1}\tan^{-1}\left(\frac{1-x}{1+x}.\sqrt{-1}\right).$$

Many of these transformations are hardly ever used in *operation*: but unless the student has them before his mind, he will be often at a loss to see the connexion of results which stand in the closest relation.

The multiplicity of value of $\log x$ (or rather λx, which might have been used throughout, as in page 47) is closely connected with that of $\sin^{-1}x$, &c. But the connexion was not very soon

---

* *Exponential;* for the logarithm is only the inverse function of the exponential one.

noticed: and the following mode of investigating the series for $\tan^{-1}x$ was consequently faulty.

Take the logarithm of both sides of $\epsilon^{2\theta\sqrt{-1}} = \dfrac{1 + \tan\theta.\sqrt{-1}}{1 - \tan\theta.\sqrt{-1}}$,

$2\theta\sqrt{-1} = 2\{\tan\theta\sqrt{-1} + \frac{1}{3}(\tan\theta\sqrt{-1})^3 + \frac{1}{5}(\tan\theta.\sqrt{-1})^5 + \ldots\}$, (*Alg.* p. 226),

or $\theta = \tan\theta - \frac{1}{3}\tan^3\theta + \frac{1}{5}\tan^5\theta - \ldots,$

a result evidently absurd, for while the first side increases from 0 to ∞, the second side goes through recurring periods. For instance, taking periods of convergency, while θ passes from $2\pi - \frac{1}{4}\pi$ to $2\pi + \frac{1}{4}\pi$, the series repeats itself for the period during which θ passes from $-\frac{1}{4}\pi$ to $\frac{1}{4}\pi$. The error lies here, $2\{\tan\theta.\sqrt{-1} + \ldots\}$ is not any logarithm we please of $\epsilon^{2\theta\sqrt{-1}}$, but some one logarithm; some one case of $2\theta\sqrt{-1} + 2m\pi\sqrt{-1}$. When we say $A = B$, therefore $LA = LB$ (page 48) we are correct only (except in the case of one isolated relation between the real and imaginary parts of A and B) on the supposition that we take the same *system*, and pair the proper logarithms of A and B in that system. And the equation $A = B$ only gives (any given logarithim of A) = (the proper logarithm of B in the same system). If we do not know the proper logarithm of B, we must take the general case, and let the conditions of the problem determine its specific meaning. Accordingly, instead of $2\theta\sqrt{-1}$, we must write $2\theta\sqrt{-1} + 2m\pi\sqrt{-1}$, and thus we have

$$\theta + m\pi = \tan\theta - \tfrac{1}{3}\tan^3\theta + \tfrac{1}{5}\tan^5\theta - \ldots\ldots,$$

which can be made true; and m must be such integer, positive or negative, as will make $\theta + m\pi$ fall between $-\frac{1}{2}\pi$ and $+\frac{1}{2}\pi$.

The following investigation requires one theorem from the theory of equations: and the rest of this chapter, generally, supposes a student who has read more than is supposed in what precedes.

Multiply together the two factors $x - \epsilon^{\theta\sqrt{-1}}$ and $x - \epsilon^{-\theta\sqrt{-1}}$; we have $x^2 - 2\cos\theta.x + 1$. Or thus, $x - \cos\theta - \sin\theta.\sqrt{-1}$ and $x - \cos\theta + \sin\theta.\sqrt{-1}$, give the product $(x - \cos\theta)^2 + \sin^2\theta$, or $x^2 - 2\cos\theta.x + 1$. Now observe that

$$x^n + \frac{1}{x^n} = 2\cos 2\theta, \quad \text{or } x^{2n} - 2x^n\cos 2\theta + 1 = 0,$$

is satisfied by

$$x + x^{-1} = 2\cos(2\theta \div n), \quad \text{or } x^2 - 2\cos\frac{2\theta}{n}.x + 1 = 0,$$

or the roots of the second equation are among the roots of the first. If 2θ be changed into $2\theta + 2\pi$, no change is made in the first equation, and the second becomes $x^2 - 2\cos\left(\frac{2\theta}{n} + \frac{2\pi}{n}\right).x + 1 = 0$, the roots of which are therefore among the roots of the first. Now two equations $x^2 - ax + 1 = 0$, $x^2 - bx + 1 = 0$, cannot have a root in common; for then it would be determined by $ax = bx$, or $x = 0$ would give the root: and there can be no such root. If therefore we go on, until we have obtained n equations of the second degree, we have got $2n$ distinct roots for the first of all. Consequently, by the theory of equations, we have, for all values of x,

$$x^{2n} - 2\cos 2\theta.x^n + 1$$

$$= \left\{x^2 - 2\cos\frac{2\theta}{n}.x + 1\right\}\left\{x^2 - 2\cos\left(\frac{2\theta}{n} + \frac{2\pi}{n}\right).x + 1\right\}\left\{x^2 - 2\cos\left(\frac{2\theta}{n} + \frac{4\pi}{n}\right)x.+1\right\}$$

$$\ldots\ldots \left\{x^2 - 2\cos\left(\frac{2\theta}{n} + \frac{(n-1)2\pi}{n}\right).x + 1\right\}.$$

When $x = 1$, $x^2 - 2\cos\phi.x + 1$ is $2(1 - \cos\phi)$ or $4\sin^2\frac{\phi}{2}$, so that

$$4\sin^2\theta = 4\sin^2\frac{\theta}{n}.4\sin^2\left(\frac{\theta}{n} + \frac{\pi}{n}\right).4\sin^2\left(\frac{\theta}{n} + \frac{2\pi}{n}\right)\ldots 4\sin^2\left(\frac{\theta}{n} + \frac{(n-1)\pi}{n}\right).$$

Extract the square root of both sides, and divide by 2;

$$\sin\theta = 2^{n-1}\sin\frac{\theta}{n}\sin\left(\frac{\theta}{n} + \frac{\pi}{n}\right)\sin\left(\frac{\theta}{n} + \frac{2\pi}{n}\right)\ldots\sin\left(\frac{\theta}{n} + \frac{(n-1)\pi}{n}\right).$$

On the second side there are n sines, or, exclusive of the first, $n - 1$ sines. If $n - 1$ be an even number we may pair these, the first and last, the second and last but one, &c. But if $n - 1$ be odd, this pairing will leave one in the middle, and $n - 1$ being odd, the middle number is $\frac{1}{2}n$, whence the middle factor is $\sin\left(\frac{\theta}{n} + \frac{n\pi}{2n}\right)$ or $\cos\frac{\theta}{n}$, which, observe, approaches unity as θ is diminished without limit. Moreover, the last one is

$$\sin\left\{\pi - \left(\frac{\pi}{n} - \frac{\theta}{n}\right)\right\} \quad \text{or} \quad \sin\left(\frac{\pi}{n} - \frac{\theta}{n}\right):$$

the last but one is, similarly, $\sin\left(\frac{2\pi}{n} - \frac{\theta}{n}\right)$, &c. Hence the pairing just alluded to gives

$$\sin\theta = 2^{n-1}\sin\frac{\theta}{n}\sin\left(\frac{\pi}{n}+\frac{\theta}{n}\right)\sin\left(\frac{\pi}{n}-\frac{\theta}{n}\right)\sin\left(\frac{2\pi}{n}+\frac{\theta}{n}\right)\sin\left(\frac{2\pi}{n}-\frac{\theta}{n}\right)\ldots$$

n factors in all, the last, a single one, being $\cos\frac{\theta}{n}$, if $n-1$ be odd. Divide both sides by $\sin(\theta\div n)$, and diminish θ without limit. The limit of the first side is then that of

$$\sin\theta\div\sin\frac{\theta}{n} \quad\text{or}\quad n\times(\sin\theta\div\theta)\div\left(\sin\frac{\theta}{n}\div\frac{\theta}{n}\right) \quad\text{or}\quad n\times 1\div 1,$$

$$n = 2^{n-1}\sin^2\frac{\pi}{n}\,.\,\sin^2\frac{2\pi}{n}\,.\,\sin^2\frac{3\pi}{n}\ \ldots\ldots \text{ one for each pair,}$$

and $\cos(\theta\div n)$, if there, has the limit unity. Now observe that

$$\sin(a+b)\sin(a-b) = 2\sin\frac{a+b}{2}\cos\frac{a-b}{2}\,.\,2\sin\frac{a-b}{2}\cos\frac{a+b}{2}$$

$$= (\sin a+\sin b)(\sin a-\sin b) = \sin^2 a - \sin^2 b.$$

Substitute from this theorem, divide $\sin\theta$ above by n, and we have, dividing both sides by θ, and transferring n,

$$\frac{\sin\theta}{\theta} = \frac{\sin\frac{\theta}{n}}{\frac{\theta}{n}}\left\{1-\frac{\sin^2\frac{\theta}{n}}{\sin^2\frac{\pi}{n}}\right\}\left\{1-\frac{\sin^2\frac{\theta}{n}}{\sin^2\frac{2\pi}{n}}\right\}\left\{1-\frac{\sin^2\frac{\theta}{n}}{\sin^2\frac{3\pi}{n}}\right\}\ldots,$$

with the factor $\cos(\theta\div n)$ at the end if $n-1$ be odd. This second side is always $\sin\theta\div\theta$, however great n may be. If we increase n without limit,

$$\frac{\sin^2\frac{\theta}{n}}{\sin^2\frac{k\pi}{n}} \quad\text{or}\quad \frac{\sin^2\left(\theta\,.\,\frac{1}{n}\right)}{\sin^2\left(k\pi\,.\,\frac{1}{n}\right)} \quad\text{has the limit}\quad \frac{\theta^2}{k^2\pi^2},$$

(see p. 18). If then we increase n without limit, we have

$$\frac{\sin\theta}{\theta} = \left(1-\frac{\theta^2}{\pi^2}\right)\left(1-\frac{\theta^2}{4\pi^2}\right)\left(1-\frac{\theta^2}{9\pi^2}\right)\left(1-\frac{\theta^2}{16\pi^2}\right)\ldots\ \textit{ad. inf.}$$

This is a remarkable converging product from which $\sin\theta$ might be calculated. Whatever θ may be, the successive factors approach to unity, and therefore produce less and less effect. A large number of factors will give a close approximation. Let $\theta = k\pi$, and we have the convenient form

$$\sin(k\pi) = k\pi\,(1-k^2)\left(1-\frac{k^2}{4}\right)\left(1-\frac{k^2}{9}\right)\left(1-\frac{k^2}{16}\right)\ldots.$$

If k be $=\frac{1}{2}$,

$$1=\frac{\pi}{2}\cdot\frac{3}{4}\cdot\frac{15}{16}\cdot\frac{35}{36}\cdot\frac{63}{64}\cdots \quad \text{or} \quad \frac{\pi}{2}=\frac{4}{3}\cdot\frac{16}{15}\cdot\frac{36}{35}\cdot\frac{64}{63}\cdot\frac{100}{99}\cdots$$

This was once suggested as a mode of approximating to the value of π; it proceeds too slowly for that purpose, but it answers another. If we take n factors, we see in the numerator the square of the product of the first n even numbers, but *not* the corresponding square of the product of odd numbers in the denominator. One odd number more is repeated once: thus in the denominator, taking 3 factors, we have 3.15.35 or $1^2.3^2.5^2.7$. Accordingly, the larger n is made, the more nearly is this equation true,

$$\left(\frac{2.4.6\ldots\ 2n}{1.3.5\ldots 2n-1}\right)^2=(2n+1)\,\frac{\pi}{2}; \quad \frac{2.4\ldots\ 2n}{1.3\ldots 2n-1}=\sqrt{(n\pi)},$$

for $(n+\frac{1}{2})\,\pi$ and $n\pi$ have nearly the same square roots, if n be very great. This, in common language, is true if n be infinite: I mean that it may be made as nearly true as we please, if n be large enough.

If this last equation were absolutely true, this next one would follow, as I shall show,

$$1.2.3\ldots n=\sqrt{(2n\pi)}\,.\left(\frac{n}{\epsilon}\right)^n.$$

But as the premise only approaches to truth as n increases, so it is also with the conclusion. Assume $1.2.3\ldots n=n^n\phi n$, or let ϕn be $1.2.3\ldots n\div n^n$.

We have then $1.2.3.4\ \ldots\ 2n-1\,.\,2n=(2n)^{2n}\,\phi\,(2n)$,

$$2\,.\,4\ \ldots\ldots\ldots\ 2n=2^n n^n\,\phi n.$$

Dividing, $\qquad 1.3.5\ \ldots\ldots\ 2n-1=(2n)^n\,\dfrac{\phi\,(2n)}{\phi n}.$

Dividing again, $\quad \dfrac{2.4.6\ \ldots\ 2n}{1.3.5\ \ldots\ 2n-1}=\dfrac{(\phi n)^2}{\phi\,(2n)}=\sqrt{(n\pi)}.$

Hence we get $\quad \dfrac{\phi\,(2n)}{\sqrt{(2\pi\,.\,2n)}}=\dfrac{(\phi n)^2}{2\pi n} \quad$ or $f(2n)=(fn)^2$,

fn standing for $\phi n\div\sqrt{(2\pi n)}$. Let $(fn)^{\frac{1}{n}}$ be ψn,

then $fn=(\psi n)^n$ and $(\psi\,2n)^{2n}=(\psi n)^{2n}$ or $\psi\,2n=\psi n$.

Consequently, ψn is either a constant, or if a function of n, does not change when n is changed into $2n$. Such functions there are, for instance, $\cos\left(2\pi\frac{\log n}{\log 2}\right)$, or any function of it except an inverse trigonometrical one. If, however, n be very great, $\log n$ increases very little when n is increased by a few units. For and about one large value of n, ψn is then nearly a constant; and, assuming it constant, we shall be able to show that the way of determining that constant gives the same thing, whatever the value of n may be. But this assumption of ψn = constant, say c, renders the proof imperfect, and a more perfect one is beyond those who may be expected to read this work. Take $\psi n = c$, then

$$fn = c^n, \quad \phi n = \sqrt{(2\pi n)} \cdot c^n, \quad \text{and } 1.2.3 \ldots n = \sqrt{(2\pi n)} \cdot c^n n^n.$$

For n write $n+1$, and divide by the former result, which gives

$$n+1 = \frac{\sqrt{\{2\pi(n+1)\}}}{\sqrt{(2\pi n)}} \frac{c^{n+1}}{c^n} \frac{(n+1)^{n+1}}{n^n} \quad \text{or} \quad 1 = \sqrt{\left(1+\frac{1}{n}\right)} \cdot c \cdot \left(1+\frac{1}{n}\right)^n.$$

If n be very great, this would give $1 = 1 \times c \times \epsilon$ very nearly (Algebra, p. 225), or $c = \epsilon^{-1}$; so that

$$1.2.3 \ldots n = \sqrt{(2\pi n)} \cdot \epsilon^{-n} \cdot n^n \text{ nearly, as asserted.}$$

This formula succeeds very well, on trial, and the first side is found greater than the second in about the proportion of $12n+1$ to $12n$.

Returning now to the form $\sin k\pi = k\pi\,(1-k^2)\ldots$, for k write $2k$, and we have

$$2\sin k\pi \cdot \cos k\pi = 2k\pi\,(1-4k^2)\left(1-\frac{4k^2}{4}\right)\left(1-\frac{4k^2}{9}\right)\left(1-\frac{4k^2}{16}\right)\left(1-\frac{4k^2}{25}\right)\ldots$$

$$\text{But} \quad \sin k\pi = k\pi \qquad \left(1-\frac{4k^2}{4}\right) \qquad \left(1-\frac{4k^2}{16}\right) \qquad \ldots$$

$$\text{or} \quad \cos k\pi = (1-4k^2)\left(1-\frac{4k^2}{9}\right)\left(1-\frac{4k^2}{25}\right)\left(1-\frac{4k^2}{49}\right)\ldots$$

a corresponding expression for the cosine. From these factorial expressions it is in our power to find series for the logarithms of trigonometrical functions. Let S_n represent the series

$$1^{-n} + 2^{-n} + 3^{-n} + 4^{-n} + \ldots \textit{ ad inf.}$$

It is easily seen that we have

$$S_n = 1^{-n} + 3^{-n} + 5^{-n} + \ldots + 2^{-n}\{1 + 2^{-n} + 3^{-n} + \ldots\}$$

$$\text{or} \quad 1^{-n} + 3^{-n} + 5^{-n} + 7^{-n} + \ldots = (1-2^{-n})\,S_n.$$

Values of S_n to a sufficient extent may be found in my *Differential Calculus*, p. 554. Now

$$\log \sin k\pi = \log k\pi + \log(1-k^2) + \log\left(1-\frac{k^2}{4}\right) + \ldots$$

Expand the several logarithms, after the first, and we have

$$\log \sin k\pi = \log k\pi - S_2 k^2 - S_4 \frac{k^4}{2} - S_6 \frac{k^6}{3} - S_8 \frac{k^8}{4} - \ldots,$$

which is convergent when $k < 1$, and very convergent when $k < \frac{1}{2}$, and, for purposes of calculation, it need never be greater. Apply the same method to the expression for $\cos k\pi$, and we have

$$\log \cos k\pi = -(2^2-1) S_2 k^2 - (2^4-1) S_4 \frac{k^4}{2} - (2^6-1) S_6 \frac{k^6}{3} - \ldots,$$

which is convergent when k is less than $\frac{1}{2}$.

From these series we get

$$\log \tan k\pi = \log k\pi + (2^2-2) S_2 k^2 + (2^4-2) S_4 \frac{k^4}{2} + (2^6-2) S_6 \frac{k^6}{3} + \ldots$$

It will be observed, that $S_2 - 1$, $S_4 - 1$, &c. diminish very rapidly. We take advantage of this by throwing the series into the form

$$\log \sin k\pi = \log k\pi + \log(1-k^2) - (S_2-1) k^2 - (S_4-1) \frac{k^4}{2} - (S_6-1) \frac{k^6}{3} - \ldots$$

$$\log \cos k\pi = -\log(1-k^2) + \log(1-4k^2) - (2^2-1)(S_2-1)k^2 - (2^4-1)(S_4-1)\frac{k^4}{4} - \ldots$$

$$\log \tan k\pi = \log k\pi + 2\log(1-k^2) - \log(1-4k^2) + (2^2-2)(S_2-1)k^2 + \ldots$$

We have seen that trigonometrical language affords a brief mode of expressing, in language derived from obvious geometrical ideas, complicated algebraical relations. The following is a striking instance:

$$\sin x = 2\cos\frac{x}{2}\sin\frac{x}{2} = 2\cos\frac{x}{2}\left(2\cos\frac{x}{4}\left(2\cos\frac{x}{8}\ldots\left(2\cos\frac{x}{2^n}\sin\frac{x}{2^n}\right)\right)\right)\ldots$$

$$\sin x - 2^n \sin\frac{x}{2^n} = \cos\frac{x}{2}\cos\frac{x}{4}\cos\frac{x}{8}\cos\frac{x}{16}\ldots\cos\frac{x}{2^n}.$$

Increase n without limit, and

$$\frac{\sin x}{x} = \cos\frac{x}{2}\cos\frac{x}{4}\cos\frac{x}{8}\cos\frac{x}{16}\cos\frac{x}{32}\ldots \textit{ad inf.}$$

Let $x = \frac{1}{2}\pi$, and show that this is then only an abbreviated form of the following,

$$\frac{2}{\pi} = \frac{\sqrt{2}}{2} \cdot \frac{\sqrt{(2+\sqrt{2})}}{2} \cdot \frac{\sqrt{(2+\sqrt{(2+\sqrt{2})})}}{2} \cdot \frac{\sqrt{(2+\sqrt{(2+\sqrt{(2+\sqrt{2})})})}}{2} \ldots$$

The student who is acquainted with the theory of equations may be enabled to express the logarithmic series in another form. The rest of this chapter is briefly given, and may be looked on as a succession of exercises. It appears from

$$\sin k\pi = k\pi (1 - k^2)(1 - \tfrac{1}{4}k^2)\ldots$$

that $\sin k\pi$ is formed from its radical factors after the manner of an algebraical expression, so that $1 - \frac{k^2}{2.3} + \ldots = 0$ may be considered as an equation of infinite dimension whose roots are $+\pi$, $-\pi$, $+2\pi$, -2π, &c. Write k for k^2, and we have in

$$1 - \frac{k}{2.3} + \frac{k^2}{2.3.4.5} - \frac{k^3}{2.3.4.5.6.7} + \ldots\ldots$$

an equation whose roots are $\pi^2, (2\pi)^2, (3\pi)^2, \ldots$ Hence we easily get the following theorems:

$$\Sigma\frac{1}{n^2} = \frac{\pi^2}{2.3}, \quad \Sigma\frac{1}{m^2.n^2} = \frac{\pi^4}{2.3.4.5}, \quad \Sigma\frac{1}{l^2.m^2.n^2} = \frac{\pi^6}{2.3.4.5.6.7}, \text{ \&c.}$$

where in $\Sigma(1 \div l^2.m^2.n^2)$, we understand that there is a term of the series for every possible combination of a product of three different integers. And by the known theorem for the reciprocals of powers of roots of an equation, we have, V_n standing for $\pi^{-2n} + (2\pi)^{-2n} + (3\pi)^{-2n} + \ldots$

$$V_1 - \frac{1}{2.3} = 0, \quad V_2 - \frac{V_1}{2.3} + \frac{2}{2.3.4.5} = 0, \quad V_3 - \frac{V_2}{2.3} + \frac{V_1}{2\ldots5} - \frac{3}{2\ldots7} = 0,$$

and so on. Calculation of a few of these results will give

$$V_1 = \frac{2}{1.2}\cdot\frac{1}{6}, \quad V_2 = \frac{2^3}{2.3.4}\cdot\frac{1}{30}, \quad V_3 = \frac{2^5}{2.3.4.5.6}\cdot\frac{1}{42},$$

and so on. Now V_n is $S_{2n}\pi^{-2n}$ (page 62), and the fractions $\frac{1}{6}, \frac{1}{30}, \frac{1}{42}$, belong to a set which are called *Bernoulli's numbers*, and are denoted by B_1, B_3, B_5, &c., so that

$$S_{2n} = \frac{1}{2}\frac{(2\pi)^{2n}}{1.2.3\ldots2n}B_{2n-1}$$

$B_1 = \frac{1}{6}$, $B_3 = \frac{1}{30}$, $B_5 = \frac{1}{42}$, $B_7 = \frac{1}{30}$, $B_9 = \frac{5}{66}$, $B_{11} = \frac{691}{2730}$, $B_{13} = \frac{7}{6}$, all which will be found by continuing the above process. These numbers appear to follow no law, which exhibited as rational fractions; but when exhibited under a law, as in

$$B_{2n-1} = 2\,\frac{1.2.3\ldots2n}{(2\pi)^{2n}}\left\{1 + \frac{1}{2^{2n}} + \frac{1}{3^{2n}} + \frac{1}{4^{2n}} + \ldots\right\},$$

it would be thought very unlikely that they should be rational fractions.

Substitution, and writing x for $k\pi$, now gives

$$\log\sin x = \log x - \frac{2B_1}{1.2}x^2 - \frac{2^3B_3}{1.2.3.4}\frac{x^4}{2} - \frac{2^5B_5}{1.2.3.4.5.6}\frac{x^6}{3} - \dots$$

$$\log\cos x = -\frac{(2^2-1)\,2B_1}{1.2}x^2 - \frac{(2^4-1)2^3B_3}{1.2.3.4}\frac{x^4}{2} - \frac{(2^6-1)2^5B_5}{1.2.3.4.5.6}\frac{x^6}{3} - \dots$$

$$\log\tan x = \log x + \frac{(2^2-2)2B_1}{1.2}x^2 + \frac{(2^4-2)2^3B_3}{1.2.3.4}\frac{x^4}{2} + \frac{(2^6-2)2^5B_5}{1.2.3.4.5.6}\frac{x^6}{3} + \dots$$

Next, we have

$$\log\cos(x+h) - \log\cos x = \log(\cos h - \tan x \,.\, \sin h)$$
$$= \log\cos h + \log(1 - \tan x \,.\, \tan h),$$
$$\log\sin(x+h) - \log\sin x = \log(\cos h + \cot x \,.\, \sin h)$$
$$= \log\cos h + \log(1 + \cot x \,.\, \tan h).$$

From the series it is obvious that $\log\cos h \div h$ diminishes without limit with h. Also it is easily deduced from page 17, that

$$\log(1 + P\tan h) \div h, \quad \text{or} \quad P(\tan h \div h) - (P^2\tan^2 h \div 2h) + \dots$$

has P for its limit when h is diminished without limit. Hence, dividing the preceding equations by h, the limits are $-\tan x$ and $\cot x$. Perform the same process on each term of the series for $\log\cos x$ and $\log\sin x$: that is, change x into $x+h$; subtract the term unaltered, and divide by h, retaining only the limit; and thus deduce the equations

$$\cot x = \frac{1}{x} - \frac{2^2B_1x}{1.2} - \frac{2^4B_3x^3}{1.2.3.4} - \frac{2^6B_5x^5}{1.2.3.4.5.6} - \dots$$
$$= \frac{1}{x} - \frac{x}{3} - \frac{x^3}{45} - \frac{2x^5}{945} - \frac{x^7}{4725} - \frac{2x^9}{93555} - \dots$$

$$\tan x = \frac{(2^2-1)\,2^2B_1x}{1.2} + \frac{(2^4-1)\,2^4B_3x^3}{1.2.3.4} + \frac{(2^6-1)\,2^6B_5x^5}{1.2.3.4.5.6} + \dots$$
$$= x + \frac{x^3}{3} + \frac{2x^5}{15} + \frac{17x^7}{315} + \frac{62x^9}{2835} + \dots$$

Take $$\cot x = \sqrt{-1}\,\frac{\epsilon^{x\sqrt{-1}} + \epsilon^{-x\sqrt{-1}}}{\epsilon^{x\sqrt{-1}} - \epsilon^{-x\sqrt{-1}}} = \sqrt{-1}\left(1 + \frac{2}{\epsilon^{2x\sqrt{-1}} - 1}\right),$$

and thence, writing $-x\sqrt{-1}$ for x, show that

$$\frac{1}{\epsilon^x - 1} = \frac{1}{x} - \frac{1}{2} + \frac{B_1x}{2} - \frac{B_3x^3}{2.3.4} + \frac{B_5x^5}{2.3.4.5.6} - \dots$$

CHAPTER VI.

ON THE CONNEXION OF COMMON AND HYPERBOLIC TRIGONOMETRY.

The system of trigonometry, from the moment that $\sqrt{-1}$ is introduced, always presents an incomplete and one-sided appearance, unless the student have in his mind for comparison (though it is rarely or never wanted for what is called use), *another system in which the there-called sines and cosines are real algebraical quantities.* This other system will serve to explain the connexion between logarithmic and trigonometrical functions.

In the ordinary system, a given revolving line, of a unit length, has one extremity in a circle; and on that circle every radius has its projections connected by the equation $x^2 + y^2 = 1$. Suppose we take all possible points so placed that the projections of their values of r are connected by the equation $x^2 - y^2 = 1$. Those points are all that are in a curve of the following form, called the *equilateral hyperbola,* (a curve corresponding, among hyperbolas, to the circle among ellipses; in fact the circle ought to be called

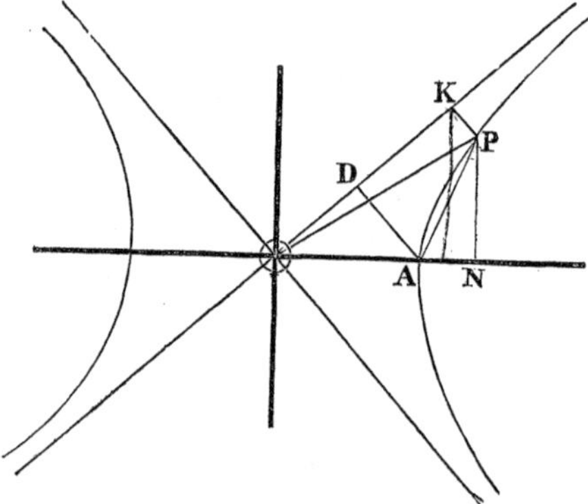

the *equilateral ellipse*). The two lines towards which the branches of the curve approach without end, but which they never meet (called *asymptotes*), are at right angles to each other, and midway between the principal axes. From any point P draw PK perpendicular to an asymptote. Let $OK = v$, $KP = w$. Then it is easily seen that $x = v \cos 45° + w \cos 45° = \frac{1}{2}\sqrt{2}\,(v + w)$, and that $y = v \sin 45° - w \sin 45° = \frac{1}{2}\sqrt{2}\,(v - w)$: whence

$$\tfrac{1}{2}(v + w)^2 - \tfrac{1}{2}(v - w)^2 = 1 \text{ or } 2vw = 1.$$

Now take one of the asymptotes and the curve that falls above it, and take two portions of the area standing on bases which are to one another as their initial distances from the centre; that is, let $OK : KL :: OK' : K'L'$. Divide each of the bases KL, $K'L'$ into n equal parts, and draw perpendiculars and inscribed rectangles in the manner shewn in the figure.

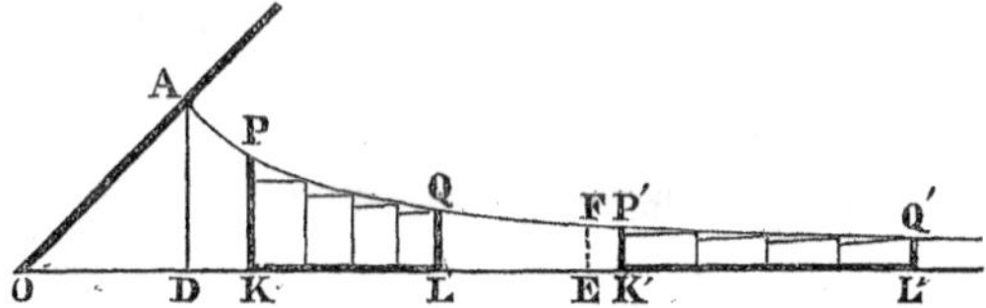

Let $OK = v$, $KP = w$, $KL = t$; each subdivision is $\frac{t}{n}$, the mth subdivision ends at $v + \frac{mt}{n}$ from the centre, so that the altitude of the mth rectangle is $1 \div 2\left(v + \frac{mt}{n}\right)$, and the area of the mth rectangle is

$$\frac{\frac{t}{n}}{2\left(v + \frac{mt}{n}\right)}, \quad \text{or} \quad \frac{1}{2\left(n\frac{v}{t} + m\right)}.$$

But in the second area $\frac{v}{t}$ is the same as in the first: therefore the mth rectangle of the first is equal to that of the second; and the sum of all the rectangles of the first is equal to the sum of all the rectangles of the second. Now the area $KPQL$ is composed of the rectangles, and of curvilinear triangles: these last, put together, fall short of a rectangle having the subdivision of KL for its base, and the fixed excess of KP over LQ for its altitude. Therefore, as the subdivision diminishes without limit, the sum of the curvilinear triangles diminishes without limit: that is, the curvilinear area is the limit of the sum of the rectangles. And as the limits of equal quantities are equal, the curvilinear areas $KPQL$ and $K'P'Q'L'$ are equal. The area $KPQL$, then, depends only on the ratio of OL and OK.

Next, OK being v, let the area $APDK$ (A being the vertex in the first figure) be A, and let $v = \phi(A)$. Put on an area $QLEF$ equal to $ADKP$ (or A) and let $ADLQ$ be B; so that $ADEF$ is $A + B$. Then, $ADKP$ and $QLEF$ being equal, we have $OD : OK :: OL : OE$.

Let $OD = m$, and this is $m : \phi A :: \phi B : \phi(A + B)$, or

$$\frac{\phi(A+B)}{m} = \frac{\phi A}{m} \cdot \frac{\phi B}{m}, \text{ whence (Algebra p. 204), } \phi A = mc^A;$$

or $v = mc^A$, whence c is to be determined. Also, $OA = 1$, and $\angle AOD = 45^\circ$, whence $m = \frac{1}{2}\sqrt{2}$. Hence $A = (\log v - \log \frac{1}{2}\sqrt{2}) \div \log c$. To determine c, observe that if we increase v by h, the increase of the area consists of a rectangle and a curvilinear triangle; and, h diminishing without limit, the ratio of the curvilinear triangle to the rectangle diminishes without limit. So that the ratio (increase of curvilinear area $\div$ inscribed rectangle) has the limit unity. Now the increase of the area is

$$\frac{\log(v+h) - \log\frac{1}{2}\sqrt{2}}{\log c} - \frac{\log v - \log\frac{1}{2}\sqrt{2}}{\log c}, \quad \text{or} \quad \frac{1}{\log c}\log\left(1 + \frac{h}{v}\right);$$

and the area of the rectangle is $h\,\dfrac{1}{2(v+h)}$;

the ratio therefore is $\dfrac{2(v+h)}{h} \cdot \dfrac{1}{\log c}\left(\dfrac{h}{v} - \dfrac{1}{2}\dfrac{h^2}{v^2} + \ldots\right)$,

and the limit of this is $2 \div \log c$, which must be 1; therefore $\log c = 2$. Accordingly $A = \frac{1}{2}\log(\sqrt{2}.v)$. The logarithms used are, as is always supposed when nothing to the contrary is mentioned, the Naperian, or *hyperbolic* logarithms: and they got the second name from their connexion with the hyperbola, the fact that all other systems are equally connected with the hyperbola not being seen when the name was given.

We can now find the curvilinear area APN. The area $DKPNA$ is made up of the rectilinear areas $DKPA$ and APN, and is therefore $\frac{1}{2}DK(DA + KP) + \frac{1}{2}AN.PN$;

or
$$\tfrac{1}{2}(v - \tfrac{1}{2}\sqrt{2})(\tfrac{1}{2}\sqrt{2} + w) + \tfrac{1}{2}(x-1)y$$
$$= \tfrac{1}{2}vw + \tfrac{1}{4}\sqrt{2}(v - w) - \tfrac{1}{4} + \tfrac{1}{2}xy - \tfrac{1}{2}y = \tfrac{1}{2}xy:$$

since $2vw = 1$, and $\frac{1}{2}\sqrt{2}(v - w) = y$. That is, the rectilinear areas ONP, $DKPNA$ are equal; take from both the curvilinear area APN, and there remains the sector APO, and the area $DKPA$ $\{\frac{1}{2}\log(\sqrt{2}.v)\}$, equal. Call the former S, and we have

$$S = \tfrac{1}{2}\log(\sqrt{2}.v) = \tfrac{1}{2}\log(x + y) = \tfrac{1}{2}\log\{x + \sqrt{(x^2 - 1)}\}.$$

Accordingly we have $x + y = \epsilon^{2S}$, $x - y = \epsilon^{-2S}$

$$x = \frac{\epsilon^{2S} + \epsilon^{-2S}}{2}, \quad y = \frac{\epsilon^{2S} - \epsilon^{-2S}}{2}.$$

If we now turn back to the circle, and if S be the area of the sector whose angle is θ, we have, the radius being 1, $S = \frac{\theta(1)^2}{2}$, or $\theta = 2S$. But now, (r being $= 1$), $x = \cos\theta$, $y = \sin\theta$, and we have

$$x = \frac{\epsilon^{2S\sqrt{-1}} + \epsilon^{-2S\sqrt{-1}}}{2}, \quad y = \frac{\epsilon^{2S\sqrt{-1}} - \epsilon^{-2S\sqrt{-1}}}{2\sqrt{-1}}.$$

If, in the hyperbola, we choose to call the numbers representing x and y the *hyperbolic* cosine and sine of the number of square units in twice the sectorial area; we have, $2S$ being θ (which is not now derived from an angle), and the difference of system being marked by capital letters in the words sine and cosine,

$$\mathrm{Cos}\,\theta = \frac{\epsilon^{\theta} + \epsilon^{-\theta}}{2}, \quad \mathrm{Sin}\,\theta = \frac{\epsilon^{\theta} - \epsilon^{-\theta}}{2}, \quad \mathrm{Tan}\,\theta = \frac{\epsilon^{\theta} - \epsilon^{-\theta}}{\epsilon^{\theta} + \epsilon^{-\theta}}.$$

From this it may be deduced that in order to convert a formula of circular trigonometry into one of hyperbolic trigonometry, when no inverse functions enter, we have but to change $\cos\theta$ into $\mathrm{Cos}\,\theta$, and $\sin\theta$ into $\sqrt{-1}\,.\,\mathrm{Sin}\,\theta$. The following are a few of the results:—

$$\mathrm{Cos}^2\theta - \mathrm{Sin}^2\theta = 1 \qquad \mathrm{Cos}(\phi \pm \theta) = \mathrm{Cos}\,\phi\,\mathrm{Cos}\,\theta \pm \mathrm{Sin}\,\phi\,\mathrm{Sin}\,\theta$$

$$\mathrm{Cos}^2\theta + \mathrm{Sin}^2\theta = \mathrm{Cos}\,2\theta, \qquad \mathrm{Sin}(\phi \pm \theta) = \mathrm{Sin}\,\phi\,\mathrm{Cos}\,\theta \pm \mathrm{Cos}\,\phi\,\mathrm{Sin}\,\theta$$

$$\mathrm{Tan}(\phi \pm \theta) = \frac{\mathrm{Tan}\,\phi \pm \mathrm{Tan}\,\theta}{1 \pm \mathrm{Tan}\,\phi.\mathrm{Tan}\,\theta};$$

$$\mathrm{Cos}^n\theta = \frac{1}{2^{n-1}}\left\{\mathrm{Cos}\,n\theta + n\,\mathrm{Cos}(n-2)\theta + n\frac{n-1}{2}\mathrm{Cos}(n-4)\theta + \ldots\right\}$$

$$\mathrm{Sin}^n\theta = \frac{1}{2^{n-1}}\{\mathrm{Cos}\,n\theta + n\,\mathrm{Cos}(n-2)\theta + \ldots\ldots\} \quad (n \text{ even})$$

$$\mathrm{Sin}^n\theta = \frac{1}{2^{n-1}}\{\mathrm{Sin}\,n\theta + n\,\mathrm{Sin}(n-2)\theta + \ldots\ldots\} \quad (n \text{ odd})$$

This is sufficient to illustrate the analogy which exists between the two systems. The advanced student may investigate the connexion of the *conjugate hyperbola* with the trigonometry in which the fundamental equation is $\text{Sin}^2\theta - \text{Cos}^2\theta = 1$.

If we now take five independent equations from page 11, say

$$\tan\theta = \frac{\sin\theta}{\cos\theta}, \quad \tan\theta\cot\theta = 1, \quad \cos\theta\sec\theta = 1, \quad \sin\theta\,\text{cosec}\,\theta = 1,$$
$$\cos^2\theta + \sin^2\theta = 1\,;$$

it is plain that the first four may be considered as equations of definition or introduction for $\tan\theta$, $\cot\theta$, $\sec\theta$, $\text{cosec}\,\theta$; and that, speaking of its operations merely, trigonometry is the treatment of the equation $x^2 + y^2 = 1$. Now as this equation might be supposed to arise from many different sources, it may be worth while to inquire how much of what precedes is due to *this form*, and how much to the *application* of this form to the *circle*, or to *angular revolution.*

If we take the two following equations,

$$x = \frac{a^\theta + a^\theta}{2}, \quad y = \frac{a^\theta - a^{-\theta}}{2\sqrt{-1}},$$

we are not bound to either, by assuming $x^2 + y^2 = 1$: but if we take one, we *must* accept the other, as will appear on trial. And then we shall find that all the direct formulæ of trigonometry follow, as soon as we require that x and y shall take the names of $\sin\theta$ and $\cos\theta$: the inverse forms depend in some measure on the meaning of a. Let a take the form $\varepsilon^{\sqrt{-1}}$, and we then regain the application of angular revolution.

CHAPTER VII.

ON THE TRIGONOMETRICAL TABLES.

THE usual trigonometrical tables are given in conjunction with tables of logarithms; and they more frequently give logarithms only than cosines, &c. themselves. When logarithms were invented, they were called *artificial* numbers; and the originals, for which logarithms were computed, were accordingly called *natural* numbers. Thus, in speaking of a table of sines, to express that it is not the logarithms of sines which are given, but sines themselves, that table would be called a table of *natural sines;* and the logarithms of these would be called, not logarithms *of sines*, but *logarithmic* sines.*

All trigonometrical tables with which the student is likely to meet, natural or logarithmic, are constructed as follows:

1. They include only the first right angle, or from 0° to 90°. If cos 96° be wanted, - sin 6° must be found; or sin 6° in the table must have its sign changed. If cos 96° be wanted in multiplication, &c., the logarithm of sin 6° must be used, and the effect of the negative sign must be properly attended to *in the final result.*

2. The arrangement is always what may be called *semi-quadrantal:* the table goes only as far as 45°, and that for the remaining half of the right angle is seen by turning the table upside down, or reading from the bottom of the page instead of the top. There is an imitation of this in the arrangement in page 17, in which $\frac{1}{2}\sqrt{3}$, which is both cos 30° and sin 60°, is read as the former by the top and the right, as the latter by the bottom and the left. Open the table so as to get tangent

---

* This leads to confusion in the minds of students, who learn some notion of mysterious identity between a number or fraction and its logarithm; and write down ·30103 = 2. The phrase is as incorrect as *royal country* would be for *king of the country*, or *constabulary parish* for *constable of the parish.*

of 37° 15′, and there will be seen, reading from the top and downwards, tangent of 37° 15′; but reading from the bottom and upwards, cotangent of 52° 45′. It would perhaps have been better if the sines had *run on* to 90°, and then all the cosines would have been in reverse readings: but the present mode is too firmly established to be shaken.

In consulting the table *inversely*, for example, in searching for the angle which has 9·61723 for the logarithm of its sine, the student must not distinguish sine from cosine, nor tangent from cotangent, but must consider *sines and cosines* as one table, *tangents and cotangents* as one table, and must cast an eye on both, and get to 9·61723 as fast as he can. For want of this caution, some beginners will turn over page after page, until they come to 45°, and then back again, perhaps to the very page that was first opened.

3. The trigonometrical tables in use were constructed on the system described in page 18, the radius being 10^{10} or ten thousand millions. Hence the logarithm of the radius was 10, and that of most sines used 9 and a fraction, 8, 7, and even 6 occurring towards the beginning or end of some tables. This has never been altered; and the consequence is that every logarithm in the tables is *too great by ten* for us. For that which we call sin θ, is sin $\theta \div 10^{10}$ of the tables. Hence, in all cases,

$$\text{Real Logarithm} = \text{Tabular Logarithm} - 10.$$

Thus, where the tables say 9·61628, we must* take out $\overline{1}.61628$, or 9·61628 − 10: where they give 12·61628, we must take 2·61628. Some tables only increase by 10 where the characteristics are negative; and give 9 for $\overline{1}$, but do not alter 0, 1, 2, &c.

When the process is inverse, the logarithm should be made tabular before entering the table with it. Thus, for finding the angle whose sine is $\overline{2}$·41729, we should enter the table with 8·41729.

---

* Many calculators prefer to consider the actual tables as if formed upon the fiction of always avoiding negative characteristics by increasing each of them by 10; and actually use the tables, making corrections in the results. For myself, I feel assured that the student should be taught by *real logarithms*, and left to find his own way to the other practice, which I much doubt his doing.

The tables* adopt various numbers of decimal places; usually five, six, or seven. Five-figure tables are exact enough for ordinary use: they may be considered as calculated to give results within the 10th part of a minute, or 6″. Those for whom five-figure tables are not sufficient, should use seven-figure tables: the six-figure tables are best for those who have *much* to do for which five figures is hardly correct enough.

In every table we use the words *argument, interval, function,* and *difference.* The *argument* is a technical term for that with which we enter a table, and opposite to which we expect to find the value of a *function* of that argument. Thus, in one table *angles* and the *logarithms of their sines* are paired: if we have a specimen of either, and want to find the corresponding one of the other kind, that with which we enter is the argument, and the other its function. The *interval* of the tables is the difference between the successive values of the principal argument, which values are always equidistant. Usually, *one minute* is the tabular interval of angle; that is, the tables furnish trigonometrical functions (or their logarithms) for 0°, 0°1′, 0°2′...1°, 1°1′...2°...90°: which I should describe as being of the class 0° (1′) 45°; the table being really in two halves, one of which is only the reverse reading of the other. But there are tables of the following descriptions: 0° (10″) 45°, 0° (1″) 45°, 0° (1″) 3° (1′) 45°, &c.

The *differences* of a table are the successive differences of the functions belonging to the equidistant arguments. Thus, if opposite to θ, $\theta + h$, $\theta + 2h$, $\theta + 3h$, &c. we have p, q, r, s, &c., the differences are $q - p$, $r - q$, $s - r$, &c., and $q - p$ is technically called *the difference* of p, $r - q$ *the difference* of q, &c.

The use of these differences lies in what is technically called *interpolation,* which is the mode of solving this question: Given the tabular function for θ, $\theta + h$, $\theta + 2h$, &c., required the proper function for $\theta +$ a given fraction of h. If the several differences be equal, or very nearly equal, as $q - p = r - q = s - r$, &c. exactly or nearly, the differences only are wanted, and the *differences of the differences,* &c. may be neglected. In this case we may

* Of ordinary tables, Hutton's (which have gone through many editions) are the best of seven figures; Farley's, of six figures; Lalande's (reprinted by Taylor and Walton), of five figures.

H

be said to use *interpolation of the first order;* and this is all that will be wanted here.

The success of interpolation of the first order depends upon the following theorem: If ϕx be a function of x, and k a small quantity, then, for every function of x, $\phi(x+k)$ is very nearly equal to $\phi x + \phi' x . k$, where $\phi' x$ is another function of x, depending upon ϕx for its form and character. I leave the student to establish the following theorems, all *nearly* true when k is small, and the angles (or k at least) are in arcual units:

$\sin(x+k) = \sin x + \cos x . k,$	$\cos(x+k) = \cos x - \sin x . k,$
$\tan(x+k) = \tan x + \dfrac{k}{\cos^2 x},$	$\cot(x+k) = \cot x - \dfrac{k}{\sin^2 x},$
$\sec(x+k) = \sec x + \dfrac{\sin x}{\cos^2 x} . k,$	$\operatorname{cosec}(x+k) = \operatorname{cosec} x - \dfrac{\cos x}{\sin^2 x} . k.$

But the second and third are not approximately true when $\cos x$ is small, nor the fifth and sixth when $\sin x$ is small.

If k be a minute or a fraction of a minute, the angle in arcual units is sufficiently* expressed by $k \sin 1'$: and then we have $\sin(x+k) = \sin x + \cos x . \sin 1' . k$, &c.

The mode of interpolation is the same as to all tables. Say that $\phi(x+k) = \phi x + \phi' x . k$, very nearly: let h be the tabular interval, and let it be required to find $\phi(x+\mu h)$, μ being a given fraction. We have then $\phi(x+h) = \phi x + \phi' x . h$, or the tabular difference is $\phi' x . h$, very nearly. But $\phi(x+\mu h) = \phi x + \phi' x . \mu h$, or $\mu \times$ tab. diff. is to be annexed, *with its proper sign*, to ϕx. Were it not for calling attention to this theorem, which is often wanted, the interpolation might be more simply explained. Take the log. sine of 3° 18′ and 3° 19′; we find 8·76015 and 8·76234, giving tab. diff. = ·00219: the table calls it 219, implying that this number and its results are to be applied in the last places. And this difference, or one very near it, runs on. We may then con-

---

* Since small angles, *arcually expressed*, are very nearly equal to their sines, $k \times$ (arcual expression of one minute) and $k \sin 1'$, are very nearly the same, if k minutes be a small angle. In astronomical books, $n \sin 1'$, $n \sin 1''$, &c. are very common substitutes for n' n'' &c. It may be worth while just to notice that, as to trigonometrical functions, it matters nothing how the angles are expressed: sin 1 (arcual unit) and sin 57° 17′ 44″ ·8.. are the same.

sider the last places as augmenting at the rate of 219 per minute. Accordingly, for half a minute we say 120, for 7-10ths of a minute 153, and so on. And it is very important to notice that sines, tangents, secants, have positive tabular differences, while *co*-sines, *co*-tangents, *co*-secants, have *negative* ones. If we want to find log cos 81° 13′·6, we must, for the *augment* of 0′·6 to the angle, *subtract* 6-tenths of the tabular difference. Neglect of this caution causes more error to beginners than anything else in their use of the tables.

The state of the tabular difference shows what degree of accuracy the tables are prepared to yield. In a table of five places, the smallest change which the table can indicate is a unit in the fifth place of decimals. Now at sin 3° 18′, there are 219 such units in the tabular difference; and each one belongs to the 219th part of a minute, about the fourth of a second. When the answer is about 3° 18′, to be determined by its sine, the problem may be solved by five-figure tables within one 219th of a minute. Accordingly, when there is choice, it is best to go to those parts of the table at which the differences are largest; shun *cosines* for small angles, and *sines* for angles near to a right angle. Again, at the beginning of the sines, the end of the cosines, and both the beginning and end of the tangents, the differences change very rapidly, and the differences of the differences become of importance. The use of the ordinary table is generally avoided in these cases, and in the following manner.

1. Generally speaking, a table of sines with smaller interval is annexed, extending over all or great part of the first degree. And the tangents and sines are very nearly equal: up to half a degree there is no practical separation.

2. At the end of the tangents, the best way is to use the tangent of the complement, which is very small, and has very nearly the same logarithm as the sine of the complement. For instance, I want to find with accuracy to five figures the tangent of 89° 46′ ·18, using the English reprint of Lalande. This is the complement of 13′·82, the tangent of which cannot be distinguished from its sine. Looking into the second small table at the end, I find 7·60360 and 7·60674 for log sin 13′·8 and log sin 13′·9 (tabular interval 0′·1). The tabular difference is 314, belonging to 0′·1; for 0′·02 I must take 2-10ths of this, or 63, which added

to the last places of 7·60360, give 7·60423; or really, $\bar{3}$·60423. This is the real logarithm of the cotangent: that of the tangent is 2·39577. The tangent then is 248·74.

I want to find the angle whose tangent is 3174. Its logarithm is 3·50161; that of the tangent (which confound with the sine) of the complement is $\bar{4}$·49839; in the tables 6·49839. In the first of the small tables appended to the English reprint of Lalande, 6·49175 and 6·49849 (with difference 674) belong to 1′ 4″ and 1′ 5″. But our *unattained part** is 664; and 674 for the whole second gives 664 for 664 ÷ 674, or ·985 of a second. The complement of the angle required is then 1′ 4″ ·985, or the angle is 89° 58′ 55″ ·015.

Nothing but actual practice can give expertness in the use of the tables. I should recommend a student to begin by verifying some formulæ. For instance (page 29) the sum of the tangents of the three angles of a triangle is equal to their product, since tan 180° = 0. Choose three angles which make 180°, find the tangents from their logarithms, and add: add the logarithms, and find the natural number to that sum: the two results ought to agree. Choose a, b at pleasure, and calculate $(a + b) \div (ab - 1)$. Find the angles to which these three are tangents: their sum ought to be 180°.

On the construction of trigonometrical tables I shall say no more than to show the student that such a thing is possible without any impracticable amount of calculation. If tables were *now* to be constructed, methods derived from the calculus of differences, which I cannot here describe, would take the place of those which I mention. But even these last are much more easy than those to which we owe the tables in use.

If we really wanted to find the sine and cosine of one minute, which, arcually expressed, is ·0002908882, we should easily find that $\theta^7 \div 2.3.4.5.6.7$ has no significant figure in the first twenty decimal places. If twenty places were enough, the following would be quite sufficient:

---

* This is a phrase which I use for the excess of what we want to find over the nearest below it which we can find. Thus if the log. tangent of the angle I want to find be 10·37466, and the nearest underneath it in the tables is 10·37461, the unattained part is 5.

$$\cos\theta = 1 - \frac{\theta^2}{2}\left(1 - \frac{\theta^2}{12}\right), \qquad \sin\theta = \theta\left\{1 - \frac{\theta^2}{6}\left(1 - \frac{\theta^2}{20}\right)\right\}.$$

Next, we have

$$\cos\theta + \sin\theta = \sqrt{(\cos^2\theta + \sin^2\theta + 2\sin\theta\cos\theta)} = \sqrt{(1 + \sin 2\theta)},$$

$$\cos\theta - \sin\theta = \sqrt{(\cos^2\theta + \sin^2\theta - 2\sin\theta\cos\theta)} = \sqrt{(1 - \sin 2\theta)},$$

when $\theta < \frac{1}{4}\pi$, both square roots must be taken positively; and we have

$$\cos\theta = \tfrac{1}{2}\sqrt{(1 + \sin 2\theta)} + \tfrac{1}{2}\sqrt{(1 - \sin 2\theta)};\ \cos\theta = \sqrt{\tfrac{1}{2}}\,(1 + \cos 2\theta),$$

$$\sin\theta = \tfrac{1}{2}\sqrt{(1 + \sin 2\theta)} - \tfrac{1}{2}\sqrt{(1 - \sin 2\theta)};\ \sin\theta = \sqrt{\tfrac{1}{2}}\,(1 - \cos 2\theta).$$

So that, either the sine or cosine of an angle being given, both the sine and cosine of its half can be found by two extractions of the square root.

Now (page 26) we may assume that we start with the sine and cosine of 3°, 6°, 9°, fully expressed for calculation. Thus we have (proceeding as directed in page 26),

$$\begin{matrix}\cos\\ \sin\end{matrix}\ 3^\circ = \frac{(\sqrt{3} \mp 1)(\sqrt{5} - 1)}{8\sqrt{2}} \pm \frac{(\sqrt{3} \pm 1)\sqrt{(5 + \sqrt{5})}}{8}.$$

Hence the sines and cosines of all the multiples of 3° may be calculated first, as verifications of the process. Having determined $\sin 1'$ and $\cos 1'$, it is now possible, by the formulæ $\sin(x + 1') = \sin x \,.\, \cos 1' + \cos x \,.\, \sin 1'$, &c., to calculate the sines and cosines of $2'$, $3'$, &c., up to $2700'$ or $45°$; which completes the table of sines and cosines, from which the tangents, &c. may be calculated by division.

Much shorter methods might be introduced, as before remarked, from the calculus of differences. But even from common formulæ, the above labour might be considerably reduced. I leave the student to prove the following formulæ:

$$\cos(30 + \theta) = \sqrt{3}\,.\cos\theta - \cos(30° - \theta), \qquad \sec\theta = \tfrac{1}{2}\tan\left(45° - \frac{\theta}{2}\right) + \tfrac{1}{2}\cot\left(45° - \frac{\theta}{2}\right),$$

$$\sin(30° + \theta) = \cos\theta - \sin(30° - \theta), \qquad \text{cosec}\,\theta = \tfrac{1}{2}\tan\theta + \tfrac{1}{2}\cot\theta,$$

$$\tan(45 + \theta) = 2\tan 2\theta + \tan(45° - \theta), \qquad \sec\theta = \tan\theta + \tan\left(45° - \frac{\theta}{2}\right),$$

$$\text{cosec}\,\theta = \cot\theta + \tan\frac{\theta}{2}.$$

From which we gather that when all sines and cosines up to 30° are calculated, the rest can be found, the sines by simple subtraction, the cosines by *one* multiplication only: that when the tangents are found up to 45°, the rest can be found by simple addition: and that all the secants and cosecants can be found by addition only, from the tangents and cotangents.

The student may also prove the following formula, which is often cited as a mode of verifying the tables, by instances selected at hazard,

$$\cos(36° + \theta) + \cos(36° - \theta) = \cos\theta + \sin(18° + \theta) + \sin(18° - \theta).$$

CHAPTER VIII.

ON THE SOLUTION OF TRIANGLES.

THIS subject, in which (and in spherical trigonometry) trigonometry was first constituted a distinct branch of mathematics, is now of little importance in a general course of mathematics. It consists mainly in the finding of convenient formulæ for the answer to the different cases of the following question: Given some parts of a triangle, to find the rest. This is called the *solution* of a triangle. But, in truth, the method given is not a solution of the problem, but a reduction of it to *the solution of a right-angled triangle.* And the *maker of the tables* it is who solves the right-angled triangle, rather than the *user* of them. The former registers, for every acute angle which consists of an exact number of minutes, all the proportions of the sides of a right-angled triangle which has that angle for one of its angles; and thus gives all the factors necessary to convert any known side into another before unknown. The latter makes use of the register, calls *himself* the sole solver of the triangle, and learns an inaccurate conception of what he has been doing.

Let the sides of a triangle contain a, b, c linear units; and let the opposite angles, gradually measured, be A, B, C. And first, let C be a right angle. By the formation of the register just alluded to, we have

$$\frac{a}{c} = \sin A = \cos B, \qquad a = c \sin A = c \cos B,$$

$$\frac{a}{b} = \tan A = \cot B, \qquad a = b \tan A = b \cot B,$$

$$c^2 = a^2 + b^2, \qquad b = \sqrt{\{(c-a).(c+a)\}}.$$

But the following formulæ should be remembered in words.

side = hypothenuse into sine of *opposite* angle,
side = hypothenuse into cosine of *adjacent* angle,
hypothenuse = *side* by sine of *opposite* angle,
hypothenuse = *side* by cosine of *adjacent* angle,
side = other side into tangent of *opposite* angle,
side = other side by tangent of *adjacent* angle.

The following are the cases which occur, and the formulæ of solution:

Given			
a, b	$\tan B = \frac{b}{a}$,	$c = \frac{b}{\sin B} = \frac{a}{\cos B}$,	$A = 90° - B$.
c, b	$a = \sqrt{(c-b.c+b)}$,	$\sin B = \frac{b}{c}$,	$A = 90° - B$.
c, A	$b = c \cos A$,	$a = c \sin A$,	$B = 90° - A$.
a, A	$b = a \cot A$,	$c = \frac{a}{\sin A}$,	$B = 90° - A$.
b, A	$a = b \tan A$,	$c = \frac{b}{\cos A}$,	$B = 90° - A$.

The following are the parts of one right-angled triangle with logarithms, for exercise in these formulæ, previously to taking other examples:

$$c = 128{\cdot}4327, \qquad \log c = 2{\cdot}1086756,$$
$$b = 66{\cdot}1364, \qquad \log b = 1{\cdot}8204405,$$
$$a = 110{\cdot}0951, \qquad \log a = 2{\cdot}0417681,$$
$$A = 59°\ 0'\ 21''{\cdot}25, \qquad B = 30°\ 59'\ 38''{\cdot}75,$$
$$\log.\sin A = \log.\cos B = \bar{1}{\cdot}9330925,$$
$$\log.\cos A = \log.\sin B = \bar{1}{\cdot}7117649,$$
$$\log.\tan A = \log.\cot B = 0{\cdot}2213276,$$
$$\log.\tan B = \log.\cot A = \bar{1}{\cdot}7786724.$$

Special cases sometimes occur in which departures from these formulæ may be advisable: as, Given b, A, where A is very small. Here $c = b \div \cos A$ is not an advantageous formula (page 75): but if we take

$$c - b = b\frac{(1-\cos A)}{\cos A} = 2b \sin^2 \frac{A}{2}, \text{ nearly,}$$

since $\cos A$ is very near unity, we get the excess of c over b very accurately.

We now proceed to triangles in general. Draw a perpendicular from the angular point of C upon c. If this perpendicular fall within the triangle, it is clear from the definition of a sine that it is $b \sin A$, and also $a \sin B$. If the perpendicular fall outside the triangle, either A or B should have its *external angle* substituted for it: but an internal and its external angle are supplements, and have the same sines. Therefore, in all cases,

$$a\sin B = b\sin A, \text{ or } \frac{a}{b} = \frac{\sin A}{\sin B}, \text{ or } \frac{a}{\sin A} = \frac{b}{\sin B}, \text{ or } a:b::\sin A:\sin B.$$

Sides, then, are to one another as the sines of their opposite angles. The angles then being given (or rather, two of them being *given*, and the third *found*), the proportions of the sides are found, being those of the sines.

I shall make this the fundamental formula from which all others are deduced, namely

$$\frac{a}{\sin A} = \frac{b}{\sin B} = \frac{c}{\sin C} \quad\ldots\ldots\ldots\ldots\ldots\ldots\ldots (1).$$

Show that each of these three, $a \div \sin A$, &c., is the *diameter of the circle circumscribing the triangle*, found in Euc. IV. 5.

The angles of a triangle being each less than two right angles, *opponents* and *completions* (page 10, note) cannot both be angles of any triangle: but *supplements* can, that is, one may be an angle of one triangle, and the other of another. When, therefore, an angle is determined by its cosine or tangent, there is but one such angle belonging to the solution: but when it is determined by its sine, there are two angles which may belong to the solution; that is, there *may* be two distinct solutions.

Now take the expanded form of $\sin(A+B)$, square both sides of the equation, and substitute values for $\cos^2 A$ and $\cos^2 B$; this gives

$$\sin^2(A+B) = \sin^2 A(1-\sin^2 B) + (1-\sin^2 A)\sin^2 B + 2\sin A\sin B\cos A\cos B$$
$$= \sin^2 A + \sin^2 B + 2\sin A\sin B\cos(A+B),$$

if A, B, C be the angles of a triangle, we have

$$A+B = 180° - C, \quad \sin(A+B) = \sin C, \quad \cos(A+B) = -\cos C;$$

whence $\sin^2 C = \sin^2 A + \sin^2 B - 2\sin A\sin B\cos C$;

divide* both sides by $\sin^2 C$, for $\sin A \div \sin C$ and $\sin B \div \sin C$, write $a \div c$, and $b \div c$, and then multiply by c^2. This gives

$$c^2 = a^2 + b^2 - 2ab\cos C \quad\ldots\ldots\ldots\ldots\ldots\ldots\ldots (2).$$

---

* This process supplies the want of a theorem with which the student ought to be acquainted in its general form. Prove that if an equation be *homogeneous* with respect to a set of letters p, q, r, &c., that equation remains true if p, q, r, &c. be erased, and p', q', r', &c. substituted, provided that p' is to p as q' to q, and as r' to r, &c.

Show that this proposition is the arithmetical representative of Euclid II. 12, 13; and that the introduction of the distinction of positive and negative quantity prevents our needing *two* propositions.

As in page 39, we may express the above thus:

$$c=(a+b)\cos\sin^{-1}\left(\frac{2\sqrt{ab}.\cos\frac{1}{2}C}{a+b}\right)=(a-b)\sec\tan^{-1}\frac{2\sqrt{ab}.\sin\frac{1}{2}C}{a-b}\ \ldots(3).$$

The formula (2) may be proved thus:—From the vertex of A draw a perpendicular upon a. In all cases it will be seen that each side of a triangle is the sum of the projections of the other two upon it, provided each projection be called positive or negative, according as the angle of projection is acute or obtuse. Thus

$$a=b\cos C+c\cos B,\quad b=c\cos A+a\cos C,\quad c=a\cos B+b\cos A.$$

Now $c^2=(c\cos A)^2+(c\sin A)^2$

$$=(b-a\cos C)^2+(a\sin C)^2=b^2-2ab\cos C+a^2.$$

The form $$\cos C=\frac{a^2+b^2-c^2}{2ab}\ \ldots\ldots\ldots\ldots\ldots\ (4)$$

is often useful. From it we have

$$1+\cos C=\frac{(a+b)^2-c^2}{2ab},\quad \cos^2\frac{C}{2}=\frac{(a+b+c)(a+b-c)}{4ab},$$

$$1-\cos C=\frac{c^2-(a-b)^2}{2ab},\quad \sin^2\frac{C}{2}=\frac{(b+c-a)(c+a-b)}{4ab}.$$

Let $a+b+c=2s$, then

$$a+b-c=2(s-c),\quad b+c-a=2(s-a),\quad c+a-b=2(s-b).$$

By substitution we have

$$\left.\begin{array}{lll}\cos^2\frac{C}{2}=\frac{s(s-c)}{ab}, & \sin^2\frac{C}{2}=\frac{(s-a)(s-b)}{ab}, & \tan^2\frac{C}{2}=\frac{(s-a)(s-b)}{s(s-c)}\\ \cos^2\frac{B}{2}=\frac{s(s-b)}{ca}, & \sin^2\frac{B}{2}=\frac{(s-c)(s-a)}{ca}, & \tan^2\frac{B}{2}=\frac{(s-c)(s-a)}{s(s-b)}\\ \cos^2\frac{A}{2}=\frac{s(s-a)}{bc}, & \sin^2\frac{A}{2}=\frac{(s-b)\sin(s-c)}{bc}, & \tan^2\frac{A}{2}=\frac{(s-b)(s-c)}{s(s-a)}\end{array}\right\}\ldots(5).$$

Let $\rho=\sqrt{\frac{(s-a)(s-b)(s-c)}{s}}$, which it will presently be shown is the radius of the inscribed circle (Euc. IV. 4).

Show that

$$\tan\frac{A}{2}=\frac{\rho}{s-a},\quad \tan\frac{B}{2}=\frac{\rho}{s-b},\quad \tan\frac{C}{2}=\frac{\rho}{s-c}\ \ldots\ldots\ldots\ldots(6);$$

$$\sin A = 2 \sin \frac{A}{2} \cos \frac{A}{2} \text{ gives } \sin A = \frac{2}{bc} \sqrt{.s(s-a)(s-b)(s-c)} \ldots (7),$$

and similar forms for $\sin B$ and $\sin C$.

If p_a, p_b, p_c, be the perpendiculars let fall upon a, b, c, from the opposite vertices, we have

$$p_a = b \sin C = c \sin B,\ p_b = c \sin A = a \sin C,\ p_c = a \sin B = b \sin A \ldots;$$

and the area of the triangle is expressed by any one of the following seven equivalent forms:

$$\frac{ap_a}{2},\ \frac{bp_b}{2},\ \frac{cp_c}{2},\ \frac{ab \sin C}{2},\ \frac{bc \sin A}{2},\ \frac{ca \sin B}{2},$$

$$\sqrt{\{s(s-a)(s-b)(s-c)\}} \text{ or } \rho s \ldots\ldots\ldots (8).$$

The formula $\frac{b}{a} = \frac{\sin B}{\sin A}$ gives $\frac{a-b}{a+b} = \frac{\sin A - \sin B}{\sin A + \sin B}$,

$$\frac{a-b}{a+b} = \frac{\tan \frac{1}{2}(A-B)}{\tan \frac{1}{2}(A+B)}, \text{ or } \tan \tfrac{1}{2}(A-B) = \frac{a-b}{a+b} \cot \frac{C}{2} \ldots (9).$$

There are four circles which touch the three *lines* of a triangle: one, the inscribed circle of Euclid, touches the three *sides*; of the others, each touches one side and the other two sides produced. Let ρ be the radius of the first, and ρ_a, ρ_b, ρ_c, those of the other three. The area of the triangle is $\frac{1}{2}(\rho a + \rho b + \rho c)$ or ρs: whence ρ, now used, is the same as ρ of the preceding formula (6). Again, the area of the triangle is $\frac{1}{2}(\rho_a b + \rho_a c - \rho_a a)$ or $(s-a)\rho_a$; whence

$$\rho_a = \frac{\sqrt{.s(s-a)(s-b)(s-c)}}{s-a}, \text{ \&c.}$$

Show that $$\frac{1}{\rho_a} + \frac{1}{\rho_b} + \frac{1}{\rho_c} = \frac{1}{\rho}.$$

Let a_b denote the projection of a on b, with its proper sign, &c. Then

$$a_b = a \cos C, \text{ \&c.},\quad b = a_b + c_b, \text{ \&c., and we have}$$

$$a_b = \frac{a^2 + b^2 - c^2}{2b},\quad b_c = \frac{b^2 + c^2 - a^2}{2c}, \text{ \&c. } \ldots\ldots (10).$$

We can now treat all the cases of oblique-angled triangles. Of two given sides or angles, let the greater of the two, when there is one, be denoted by the prior letter of the alphabet.

1. Given the three sides to find the angles.

If one angle only be wanted, say A, take

$$\cos\frac{A}{2} = \sqrt{\frac{s(s-a)}{bc}} \quad \text{or} \quad \sin\frac{A}{2} = \sqrt{\frac{(s-b)\ (s-c)}{bc}},$$

preferring the first for the greatest angle, the second for the least. If all the three angles be wanted (or even two), take

$$\rho = \sqrt{\{(s-a)(s-b)(s-c) \div s\}}$$

$$\tan\tfrac{1}{2}A = \rho \div (s-a),\ \tan\tfrac{1}{2}B = \rho \div (s-b),\ \tan\tfrac{1}{2}C = \rho \div (s-c),$$

which verify each other, since $\frac{1}{2}A + \frac{1}{2}B + \frac{1}{2}C = 90°$.

This method was once much used. Since

$$a_b^2 + p_b^2 = a^2,\quad c_b^2 + p_b^2 = c^2,\quad (a_b - c_b)(a_b + c_b) = (a-c)(a+c),$$

$$\text{or} \quad a_b - c_b = \frac{(a-c)(a+c)}{b},\quad a_b + c_b = b;$$

from which determine a_b and c_b, and then use

$$\cos C = \frac{a_b}{a},\quad \cos A = \frac{c_b}{c}.$$

2. Given two sides and the included angle (a, b, C), to find the rest. If all be wanted, calculate the angles by means of their half sum and half difference, thus,

$$\tfrac{1}{2}A + \tfrac{1}{2}B = 90° - \tfrac{1}{2}C,\quad \tan(\tfrac{1}{2}A - \tfrac{1}{2}B) = \frac{a-b}{a+b}\tan(90° - \tfrac{1}{2}C),$$

$$A = (\tfrac{1}{2}A + \tfrac{1}{2}B) + (\tfrac{1}{2}A - \tfrac{1}{2}B),\quad B = (\tfrac{1}{2}A + \tfrac{1}{2}B) - (\tfrac{1}{2}A - \tfrac{1}{2}B).$$

Lastly, $c = a\dfrac{\sin C}{\sin A} = b\dfrac{\sin C}{\sin B}$, which ought to agree.

If only the side c be wanted, take either

$$c = (a+b)\cos\sin^{-1}\frac{2\sqrt{ab}\cos\frac{1}{2}C}{a+b} = (a-b) \div \cos\tan^{-1}\frac{2\sqrt{ab}\sin\frac{1}{2}C}{a-b};$$

say, for reference,

$$(a+b)\cos S_c \text{ and } (a-b) \div \cos T_c.$$

Or thus, $\qquad b_a = b\cos C,\ p_a = b\sin C,$ which find:

$$c_a = a - b_a,\quad \tan B = \frac{p_a}{c_a},\quad c = \frac{c_a}{\cos B},$$

which is a direct reduction of the solution to that of right-angled triangles.

3. Given one side (c) and two angles, required the other sides. Calculate the third angle, and then use

$$a = c\frac{\sin A}{\sin C}, \qquad b = c\frac{\sin B}{\sin C};$$

but if C be obtuse, use $A+B$ instead of C. Or, in any case, $A+B$ may be used, taking the cosine of the excess above 90° (which excess is easily found without pen and ink) when $A+B$ is obtuse.

4. When two sides and an angle not included are given, (a, b, B), required the rest.

First calculate A from $\sin A = \dfrac{a \sin B}{b}$.

If $a \sin B > b$, $\sin A > 1$, and there is no solution. If $a \sin B = b$, then $\sin A = 1$, and A is a right angle, as is its supplement: there is this one solution only, and $c = a \cos B$, $C = 90° - B$. But when $a \sin B < b$, $\sin A < 1$, and there are two solutions, one acute and one obtuse. Let these be A' and A''; and let C' and C'', c' and c'', be the corresponding values of the third angle and side; then $C' = 180° - B - A'$, $C'' = 180° - B - A''$, and

$$c' = \frac{a \sin C'}{\sin A} = \frac{b \sin C'}{\sin B}, \qquad c'' = \frac{a \sin C''}{\sin A} = \frac{b \sin C''}{\sin B}.$$

So far it seems as if we were sure of two solutions, whenever $a \sin B < b$. But in trigonometry we are often made to observe what meets us so frequently in ordinary Algebra, namely, that in constructing the conditions of a problem, we are compelled to take in those of cognate problems. If we have not until now met with such a circumstance in this present chapter, it is because in mere solution of triangles, we have not introduced into this isolated subject our conventions as to the measurement of angles, which would, as in page 10, oblige us to consider *completions* and *opponents* of Euclid's angles as among the angles of a triangle, and each *cosine* and *tangent* as belonging to two possible angles of a triangle. But each of two *supplements* may be the angle of a triangle: and when, in our construction, we have to use the *sine* of a *given angle*, we conjoin with our problem that one in which the *supplement* of the first given angle is made the given angle. Let B' and B'' be the acute and obtuse supplements: then all we are entitled to say is that the preceding solutions belong, each of them, to one or the other of the triangles $(a,\ b,\ B')$ $(a,\ b,\ B'')$.

First, suppose the given angle to be opposite the lesser of the given sides $(b < a)$. Then $B < A$, and there cannot be any obtuse value of B: that is, both solutions belong to the acute value of B. Secondly, let the given sides be equal $(b = a)$: then $B = A$, and A must be acute, and $180° - B - A''$, or $180 - A' - A''$, is $= 0$. One of the solutions vanishes into a straight line, and the other is an isosceles triangle. Thirdly, let the given angle be opposite the greater of the given sides $(b > a)$. Then $B > A$, there cannot be any obtuse value of A, but both values of B may be used, and one solution belongs to each value. The following diagram will explain these cases.

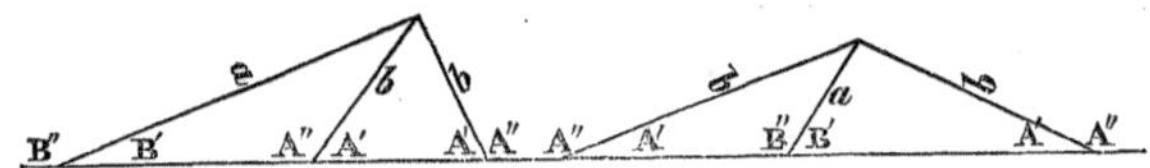

This double solution is, as might be supposed, indicative of the problem being one of the second degree. We have

$$b^2 = c^2 + a^2 - 2ac \cos B, \quad c = a \cos B \pm \sqrt{(b^2 - a^2 \sin^2 B)}.$$

Here $a \sin B$ is p_c, and $\pm \sqrt{(b^2 - a^2 \sin^2 B)}$ is b_c with its sign, on the supposition that c is measured positively on the side of the acute value of B; $a \cos B$ is a_c; and the above equation is only $c = a_c + b_c$, in which c has its proper sign. The consideration of this problem, and of its connexion with Euclid VI. 7, will be a useful exercise.

The following table shews all the parts of a triangle and their logarithms for exercise.

$a = 15{\cdot}236$ $\log a = 1{\cdot}1828710$ $s - a = 3{\cdot}098$ $\log(s - a) = 0{\cdot}4910814$

$b = 12{\cdot}414$ $\log b = 1{\cdot}0939117$ $s - b = 5{\cdot}920$ $\log(s - b) = 0{\cdot}7723217$

$c = 9{\cdot}018$ $\log c = 0{\cdot}9551102$ $s - c = 9{\cdot}316$ $\log(s - c) = 0{\cdot}9692295$

$s = 18{\cdot}334$ $\log s = 1{\cdot}2632572$ $\rho s = 55{\cdot}96866$ $\log \rho s = 1{\cdot}7479449$

$a + b = 27{\cdot}650$ $\log(a + b) = 1{\cdot}4416951$ $a - b = 2{\cdot}822$ $\log(a - b) = 0{\cdot}4505570$

$b + c = 21{\cdot}432$ $\log(b + c) = 1{\cdot}3310627$ $b - c = 3{\cdot}396$ $\log(b - c) = 0{\cdot}5309677$

$a + c = 24{\cdot}254$ $\log(a + c) = 1{\cdot}3847834$ $a - c = 6{\cdot}218$ $\log(a - c) = 0{\cdot}7936507$

		log. sin.	log. cos.	log. tan.
A	$= 89^\circ\ 9'\ 23''\cdot 54$	$\bar{1}\cdot 9999530$	$\bar{2}\cdot 1679268$	$1\cdot 8320262$
B	$= 54\ 33\ 25\cdot 12$	$\bar{1}\cdot 9109937$	$\bar{1}\cdot 7633479$	$0\cdot 1476458$
C	$= 36\ 17\ 11\cdot 48$	$\bar{1}\cdot 7721922$	$\bar{1}\cdot 9063714$	$\bar{1}\cdot 8658208$
$\frac{1}{2}A$	$= 44\ 34\ 41\cdot 77$	$\bar{1}\cdot 8462647$	$\bar{1}\cdot 8526583$	$\bar{1}\cdot 9936064$
$\frac{1}{2}B$	$= 27\ 16\ 42\cdot 56$	$\bar{1}\cdot 6611649$	$\bar{1}\cdot 9487989$	$\bar{1}\cdot 7123661$
$\frac{1}{2}C$	$= 18\ 8\ 35\cdot 74$	$\bar{1}\cdot 4933102$	$\bar{1}\cdot 9778520$	$\bar{1}\cdot 5154583$
$\frac{1}{2}(A-B)$	$= 17\ 17\ 59\cdot 21$	$\bar{1}\cdot 4722989$	$\bar{1}\cdot 9798951$	$\bar{1}\cdot 4924039$
$\frac{1}{2}(B-C)$	$= 9\ 8\ 6\cdot 82$	$\bar{1}\cdot 2007551$	$\bar{1}\cdot 9944564$	$\bar{1}\cdot 2062988$
$\frac{1}{2}(A-C)$	$= 26\ 26\ 6\cdot 03$	$\bar{1}\cdot 6485379$	$\bar{1}\cdot 9520365$	$\bar{1}\cdot 6965016$
S_a	$= 44\ 41\ 30\cdot 6$	$\bar{1}\cdot 8471366$	$\bar{1}\cdot 8518083$	....
S_b	$= 59\ 12\ 50\cdot 4$	$\bar{1}\cdot 9340361$	$\bar{1}\cdot 7091283$	....
S_c	$= 70\ 57\ 53\cdot 6$	$\bar{1}\cdot 9755783$	$\bar{1}\cdot 5134140$	....
T_a	$= 77\ 7\ 15\cdot 5$	....	$\bar{1}\cdot 3408970$	$0\cdot 6408380$
T_b	$= 59\ 56\ 28\cdot 9$	....	$\bar{1}\cdot 6997390$	$0\cdot 2375348$
T_c	$= 71\ 45\ 50\cdot 9$	....	$\bar{1}\cdot 4954464$	$0\cdot 4821746$

$p_a = 7\cdot 347 \quad \log p_a = 0\cdot 8661039$

$p_b = 9\cdot 017 \quad \log p_b = 0\cdot 9550632$

$p_c = 12\cdot 413 \quad \log p_c = 1\cdot 0938647$

$b_a = 10\cdot 006 \quad \log b_a = 1\cdot 0002831 \qquad c_a = 5\cdot 230 \quad \log c_a = 0\cdot 7184581$

$a_b = 12\cdot 281 \quad \log a_b = 1\cdot 0892424 \qquad c_b = \cdot 133 \quad \log c_b = \bar{1}\cdot 1230370$

$a_c = 8\cdot 835 \quad \log a_c = 0\cdot 9462189 \qquad b_c = \cdot 183 \quad \log b_c = \bar{1}\cdot 2618385.$

Cases may occur in which the particular values of the data render special methods convenient. For instance, when a, b, C, are given, and $C = 180^\circ - C_{,}$, $C_{,}$ being very small, we may proceed as follows:

$$c^2 = a^2 + b^2 - 2ab\cos(180^\circ - C_{,}) = a^2 + b^2 + 2ab\cos C_{,}$$
$$= (a+b)^2 - 4ab\sin^2\tfrac{1}{2}C_{,} = (a+b)^2\left\{1 - \frac{4ab}{(a+b)^2}\sin^2\tfrac{1}{2}C_{,}\right\}.$$

By the binomial theorem, $\sqrt{(1-x)} = 1 - \frac{1}{2}x$ nearly, x being small,

$$c = (a+b)\left(1 - \frac{2ab}{(a+b)^2}.\sin^2\tfrac{1}{2}C_{,}\right) = a + b - \frac{2ab}{a+b}\sin^2\tfrac{1}{2}C_{,}$$

very nearly. But $\sin C_{,} = 2\sin\frac{1}{2}C_{,}.\cos\frac{1}{2}C_{,}$, or $\cos\frac{1}{2}C_{,}$ being very nearly 1, we have

$$\sin\tfrac{1}{2}C_{,} = \tfrac{1}{2}\sin C_{,}, \quad \sin^2\tfrac{1}{2}C_{,} = \tfrac{1}{4}\sin^2 C_{,}, \text{ very nearly:}$$

$$c = a + b - \frac{1}{2}\frac{ab\sin^2 C_{,}}{a+b}, \text{ very nearly.}$$

If the circumscribed circle be drawn, the angle of the radii drawn to the extremities of a, is the angle at the centre to which A is the angle at the circumference. There is, then, an isosceles triangle, in which r and r include either $2A$, or $360° - 2A$, the third side being a. Consequently,

$$a = 2r \sin(A \text{ or } 180° - A):$$

that is, $2r = a \div \sin A$. The three sides of the triangle are then

$$2r \sin A, \quad 2r \sin B, \quad 2r \sin C;$$

and all the formulæ become trigonometrical identities, if these be substituted for a, b, c. Thus, substitution in the formulæ for $\sin^2 \frac{1}{2}C$ gives us

$$\sin^2 \frac{C}{2} = \frac{(\sin B + \sin C - \sin A)(\sin C + \sin A - \sin B)}{4 \sin A \sin B};$$

which is always true when $A + B + C = 180°$.

Shew that the line drawn from the vertex of A bisecting the *side* a is $\sqrt{\{\frac{1}{2}(b^2 + c^2) - \frac{1}{4}a^2\}}$ or $\frac{1}{2}\sqrt{\{b^2 + c^2 + 2bc \cos A\}}$. Also that the line bisecting the angle A is $2bc \cos \frac{1}{2}A \div (b + c)$. Shew that

$$16s(s - a)(s - b)(s - c) = 2b^2c^2 + 2c^2a^2 + 2a^2b^2 - a^4 - b^4 - c^4.$$

BOOK II.

DOUBLE ALGEBRA.

CHAPTER I.

DESCRIPTION OF A SYMBOLIC CALCULUS.

THE object of this book is the construction of Algebra upon a basis which will enable us to give a meaning to every symbol *and combination of symbols* before it is used, and consequently to dispense, first, with all unintelligible combination, secondly, with all search after interpretation of combinations subsequently to their first appearance.

In arithmetic and in ordinary algebra we use *symbols* of previously assigned *meaning*, from which meaning, by self-evident notions of number, &c., are derived *rules of operation*. The student must understand by *symbols*, the *peculiar* symbols of arithmetic and algebra: strictly speaking, the written or spoken words by which meaning is conveyed are themselves symbols. And symbols must be explained by other symbols, except when they denote external objects or actions, in which case the symbol may be explained by pointing to the object present or the action taking place. Language itself is a science of symbols (namely, words) having meanings (which are described in the *dictionary* by words of the same or another language) and rules of combination (laid down in its *grammar*).

No science of symbols can be fully presented to the mind, in such a state as to demand assent or dissent, until its peculiar symbols, their meanings, and the rules of operation, are *all* stated. In this case we have but to ascertain—first, whether the peculiar symbols be distinguishable from each other; secondly, whether the meanings are capable of being distinctly apprehended, each symbol having either one only, or an attainable and intelligible choice; thirdly, whether the given rules of operation be necessary consequences of the given meanings as applied to the given symbols. If these inquiries produce as many affirmative answers, the basis of the science is *so far unobjectionable;* and all intelligible conclusions which are drawn from a correct and intelligible use of the rules of operation, are true. But yet it may be *imperfect.*

First, it may be *incomplete in its peculiar symbols.* There may be a want of symbols which those already in use suggest, but which are not made to appear. This is not the incompleteness to which algebra is most liable: it suffers more from its symbolic combinations growing much faster than the ordinary language in which they are, if possible, to be occasionally expressed.

Secondly, it may be *incomplete in its meanings.* For example, it may be capable of applying, with the same symbols, to more subjects than its actual meanings take in. This is one possible incompleteness, of a very obvious character. Another, of a much less obvious character, and which probably nothing but actual experience of it would have suggested, is this: symbols, defined in a manner which makes them separately intelligible, may be unintelligible in combination; their separate definitions may involve what, in the attempt* to combine them, produces con-

* The student may be surprised at my saying that we should never have imagined such a result in algebra without actual experience of it: for it may strike him immediately that in ordinary language we may have not merely unmeaning, but contradictory, combinations. But the answer is that we are so accustomed to contradictory combinations, used in some emphatic sense, that they are recognised idioms: it even happens that they express more and better meaning while they are fresh, and before use makes the contradiction wear off, than afterwards. When General Wolfe first used the expression 'choice of difficulties', which was contradiction, choice then meaning voluntary election, he made those to whom he wrote see his position with much more effect than could have been produced a second time by the same words. Ordinary language has methods of instantaneously assigning meaning to contradictory phrases: and thus it has stronger analogies with an algebra (if there were such a thing) in which there are pre-organized rules for explaining new contradictory symbols as they arise, than with one in which a single instance of them demands an immediate revision of the whole dictionary.

tradiction. The second case may be a consequence of the first, or it may not: contradictory combinations may arise from limitation of meaning, and may cease to be contradictory under extended meanings; or it may happen, either that no such abolition of contradiction is possible in the case thought of, or else that every extension of meaning which destroys contradiction in one combination creates it in another.

Thirdly, it may be *incomplete in its rules of operation.* This incompleteness may amount either to an absolute privation of results, or only to the imposition of more trouble than, with completeness, would be requisite. Every rule the want of which would be a privation of results, may be called *primary:* all which might be dispensed with, except for the trouble that the want of them would give, may be treated merely as consequences of the primary rules, and called *secondary.*

Each of the three great objects of consideration, peculiar symbols, assigned meanings, and rules of operation, may then be defective, independently of the rest. Can we carry the defect so far as to imagine one or more of them to be entirely wanting? The cases of absolute deficiency, which it may be worth while to notice here, principally to accustom the student to the idea of the separation, are as follows:—

1. *Meanings and rules without peculiar symbols.* Unquestionably algebra *might* be deprived of its peculiar symbols, ordinary words taking their places. There is no more truth, no more meaning, and no more possibility of drawing consequences in

$$(a^2 - b^2) \div (a - b) = a + b,$$

than in 'the difference of the products of two numbers, each multiplied by itself, divided by the difference of those numbers, gives the sum for a quotient.' Before the time of Vieta, algebra had always been much retarded by the want of a sufficient use of peculiar symbols.

2. *Peculiar symbols, and meanings, without rules of operation.* In this case the only process must be one of unassisted reason, thinking on the objects which the symbols represent; as in geometry, which has its peculiar symbols (as AB, signifying a line joining two points named A and B). But no science

of *calculation** can proceed without rules; and these geometry does† not possess.

3. *Peculiar symbols, and rules of operation, without assigned meanings.* Nothing can be clearer than the possibility of dictating the symbols with which to proceed, and the mode of using them, without any information whatever on the meaning of the former, or the purpose of the latter. A corresponding process takes place in every manual art in which an assistant obeys directions, without understanding them. The use of such a process, as an exercise of mind, must depend much (but not altogether) upon the value of the meanings which we suppose are to be ultimately assigned. A person who should learn how to put together a map of Europe dissected before the paper is pasted on, would have symbols, various shaped pieces of wood, and rules of operation, directions to put them together so as to make the edges fit, and the whole form an oblong figure. Let him go on until he can do this with any degree of expertness, and he has no consciousness of having learnt anything: but paste on the engraved paper, and he is soon made sensible that he has become master of the forms and relative situations of the European countries and seas.

As soon as the idea of acquiring symbols and laws of combination, without given meaning, has become familiar, the student has the notion of what I will call a *symbolic calculus;* which, with certain symbols and certain laws of combination, is *symbolic algebra:* an art, not a science; and an apparently

---

* A *calculus*, or *science of calculation*, in the modern sense, is one which has organized processes by which passage is made, or may be made, mechanically, from one result to another. A *calculus* always contains something which it would be *possible* to do by machinery.

† Those who introduce *algebraical* symbols into elementary geometry, destroy the peculiar character of the latter to every student who has any mechanical associations connected with those symbols; that is, to every student who has previously used them in ordinary algebra. Geometrical reasoning, and arithmetical process, have each its own office: to mix the two in elementary instruction, is injurious to the proper acquisition of both.

useless art, except as it may afterwards furnish the grammar of a science. The proficient in a symbolic calculus would naturally demand a supply of meaning. Suppose him left without the power of obtaining it from without: his teacher is dead, and he must *invent meanings* for himself. His problem is, Given symbols and laws of combination, required meanings for the symbols of which the right to make those combinations shall be a logical consequence. He tries, and succeeds; he invents a set of meanings which satisfy the conditions. Has he then supplied what his teacher would have given, if he had lived? In one particular, certainly: he has turned his *symbolic* calculus into a *significant* one. But it does not follow that he has done it in the way which his teacher would have taught him, had he lived. It is possible that many* different sets of meanings may, when attached to the symbols, make the rules necessary consequences. We may try this in a small way with three symbols, and one rule of connexion. Given symbols M, N, $+$, and one sole relation of combination, namely that $M+N$ is the same result (be it of what kind soever) as $N+M$. Here is a symbolic calculus: how can it be made a significant one? In the following ways, among others. 1. M and N may be *magnitudes*, $+$ the sign of addition of the second to the first. 2. M and N may be *numbers*, and $+$ the sign of multiplying the first by the second. 3. M and N may be *lines*, and $+$ a direction to make a rectangle with the antecedent for a base, and the consequent for an altitude. 4. M and N may be *men*, and $+$ the assertion that the antecedent is the brother of the consequent. 5. M and N may be nations, and $+$ the sign of the consequent having fought a battle with the antecedent: and so on.

We may also illustrate the manner in which too limited or too extensive a meaning interferes with the formation of the most complete significant calculus. In (1), limitation to *mag-*

---

* Most inverse questions lead to multiplicity of answers. But the student does not fully expect this when he asks an inverse question, unless he be familiar with the logical character of the predicate of a proposition. A always gives B: what gives B? answer, A always, and, for aught that appears, many other things.

nitude is not necessary, unless *ratio* and *number* be signified under the term. In (2), if M (only) were allowed to signify number, $N+M$ would be intelligible, but $M+N$ would be unintelligible; an *impossible* symbol of this calculus. In (3), $(M+N)$ signifying the rectangle, $(M+N)+P$ would be unintelligible at first: further examination would show that the explanation is not complete; and that the proper *extension* is that $M+N+P$ should signify the formation of the *right solid* (rectangular parallelepiped) with the sides M, N, P. But $M+N+P+Q$ will be always unintelligible, as space has not four dimensions. In (4), the extension of M and N to signify *human beings*, would spoil the applicability of the rule, unless the meaning of + were at the same time extended to signify the assertion that the antecedent was *brother or sister* (as the case might be) of the consequent.

But when the symbols are many, and laws of combination various, is it to be thought possible such a number of coincidences should occur, as that the same symbolic combinations (unlimited in number) which express truths under one set of meanings, express other truths under another? Could two different languages be contrived, having the same words and grammar, but in which the words have different meanings, in such manner that any sentence which has a true meaning in the first, should also have a true, but a different, meaning in the second? This last question may almost certainly be answered in the negative: the thousands of arbitrary terms which a language presents, and the hundreds of grammatical junctions, present a possible variety of combinations of which it would be hopeless to expect an equal number of coincidences of the kind required. But Algebra has few symbols and few combinations, compared with a language: more explanations than one are practicable, and many more than have yet been discovered may exist. And the student, if he should hereafter inquire into the assertions of different writers, who contend for what each of them considers as *the* explanation of $\sqrt{-1}$, will do well to substitute the indefinite article.

We can now form some idea of the object in view; and we must ask, first, what are the steps through which we have gone, to arrive at algebra as it stands in the mind of the student who commences this book. They are, very briefly, as follows:

Beginning with *specific* or *particular arithmetic*, in which every symbol of number has one meaning, we have invented signs, and investigated rules of operation. An easy ascent is made to *general* or *universal arithmetic*, in which general symbols of number are invented, the letters of the alphabet being applied to stand for numbers, each letter having a numerical meaning, known or unknown, on each occasion of its use. And thus, omitting many circumstances which have no particular reference to our present subject, we arrive at a calculus in which the actual performance of computations is deferred until we come to the time when the values of the letters are found or assigned. Accordingly, whereas in particular arithmetic every computation is completed as it arises, or declared impossible, in universal arithmetic we have a calculus of forms of computation, in which each numerical computation is only signified, and not performed; the proviso, *if possible*, being annexed by a reasoner to every step of every process in which a chance of impossibility occurs.

Out of a few cases of difficulty, there is selected one, which appears at first sight destined always to make the proviso above mentioned an essential part of most processes of universal arithmetic. It is the *impossible subtraction;* the constant appearance in problems of a demand to take the greater from the less, to say how many units there are in 6 - 20, for instance. An examination of the circumstances under which such phenomena occur shows, inductively, that their producing cause is always this, that either in the statement of the problem, or in its treatment, some one quantity is supposed to be of a kind diametrically opposite to that which it ought to be.

Simple number, the subject of abstract arithmetic, be it particular or universal, fails to show any acknowledgment of a distinction which strikes us in almost every notion of concrete magnitude. Measure 10 feet from a given point on a given line: the command is ambiguous until we are told which of two directions to take. A sum of money in the concerns of A and Co. is incapable of being entered in their books until we know whether it be gain or loss. A weight is generally of one kind, but not always: the weight of a balloon is a tendency in the direction opposite to that of most weights; or rather, the word weight being by usage not allowed a double signification, we say a balloon has no weight, but something which is the direct opposite of weight. A time, one extreme epoch of which is mentioned, is not sufficiently described until we know whether it is all before, or all after, the epoch. And so on. In every one of these cases, the numerical quantity of a concrete magnitude, described by means of a standard unit, is not a sufficient description; it is necessary to specify to which of two opposite kinds it belongs. This specification must be made by something not numerical: number is wholly inadequate.

The first suggestion would be, it might be thought, to invent signs of distinction: but universal arithmetic makes a suggestion which forces attention, before the necessity for distinction is more than barely perceived. Should we ever suppose that the result of a problem is gain, or distance in one direction, or time after an epoch, &c., when it is in reality, say 4 of loss, or of distance in the opposite direction, or of time before the epoch, &c., the answer always presents itself as $0 - 4$, or $m - (m + 4)$, or as some version of the attempt to take away 4 more than there are to be taken away. It is then judged convenient (that the convenience amounts to a necessity is hardly seen at that period) to make -4 the symbol of 4 units of a kind directly opposite to those imagined in 4, or $0 + 4$. And this is the first of the steps by which universal arithmetic becomes common, or *single* algebra. See Algebra, pp. 12–19 and 44–66, for more detail.

This word *single*, as applied to algebra, is derived from space of *one* dimension, or length, in which it is always possible to represent the effect of every intelligible operation of single algebra, and the interpretation of every result which admits of any interpretation at all. When we reckon time, gain and loss, &c., it is always possible to translate our reckoning into terms of length, as follows:

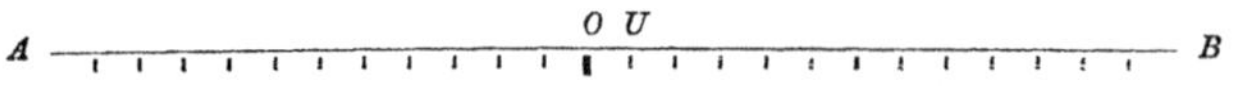

O U

A ———————————————————————— *B*

Take any point O, in a straight line, which call the *zero-point*, from which all measurement is to begin. Let OU represent the unit of any particular magnitude, and let magnitudes of one kind, say gains, be measured towards A, and losses towards B. Successive gains and losses may be taken off, and the final balance exhibited, by the compasses. As long as the result is always of one kind, so that an assumption to that effect would never render the processes of pure arithmetic unintelligible, the successive results always appear on one side of O: but the moment a result of the contrary kind appears, (which, unless the arithmetical computer were aware of it, and had provided accordingly, would leave him with an attempt at impossible subtraction on his hands,) it is indicated on the opposite side of O.

The convention as to the meaning of $+1$ and -1, namely, that they shall represent units of diametrically opposite kinds, is a very bold one: not merely because it takes up signs which are originally intended for nothing but addition and subtraction, and fixes another signification on them; but because it still employs them to connect quantities, *and by a new kind of connexion*. The signs in fact are used in two senses, the *directive*, and the *conjunctive*. $+(-3)$ tells us, by virtue of $-$, what we are talking of, and by virtue of $+(\ )$ how we are to join what we talk of to the rest. As conjunctive signs, $+$ means *junction*, or putting on what we speak of; and $-$ means *removal*. Thus, if $+$ and $-$ in the directive sense indicate gain and loss, the question, What is

$$(-3)+(+8)-(-7)+(-4)-(+3)?$$

is the following:—A man loses 3, and gets a gain of 8, with the removal of a loss of 7, the accession of a loss of 4, and the removal of a gain of 3: what is the united effect of all these actions on his previous property? The answer is, the accession of a gain of 5, $+(+5)$.

The mere beginner is allowed to slide into single algebra from universal arithmetic in a manner which leads him to underrate the magnitude of the change. I do not see that it can be otherwise: but, at this period, my reader may be made to observe that the process by which we shall pass from single to double algebra, is the surest and most demonstrative (perhaps the only demonstrative) mode of passing from universal arithmetic to single algebra. It is not until he can drop all meanings, collect the laws of combination of the symbols, and so form a purely symbolic calculus, and then proceed to furnish that calculus with extended meanings, that he becomes fully master of the change. But the close resemblances, which make the slide above referred to so easy, might make it doubtful whether he would be fit to take proper note of this case of *reduction and restoration** until he has seen a more striking form of the same process, namely, that which is exhibited in the transition from single to double algebra.

When the earlier algebraists first began to occupy themselves with questions expressed in general terms, the difficulties of subtraction soon became obvious, inasmuch as the greater would sometimes demand to be subtracted from the less. The science has been brought to its present state through three distinct steps. The first was tacitly to contend for the principle that human faculties, at the outset of any science, are judges both of the extent to which its results can be carried, and of the form in which they are to be expressed. *Ignorance*, the necessary predecessor of knowledge, was called *nature;* and all conceptions which were declared unintelligible by the former, were supposed to have been made impossible by the latter. The first who used algebraical symbols in a general sense, Vieta, concluded that subtraction was a defect, and that expressions containing it should be in every possible manner avoided. *Vitium negationis*, was his phrase. Nothing could make a more easy pillow for the mind, than the rejection of all which could give any trouble; but if Euclid had altogether dispensed with the *vitium paral-*

---

* Algebra, *al jebr e al mokābala*, restoration and reduction, got its Arabic name, I have no doubt, from the *restoration* of the term which completes the square, and *reduction* of the equation by extracting the square root. The solution of a quadratic equation was the most prominent part of the Arabic algebra. Alter the order of the words, and the phrase may well represent the final mode of establishing algebra: *reduction* of universal arithmetic to a symbolic calculus, followed by *restoration* to significance under extended meanings.

lelorum, his geometry would have been confined to twenty-six propositions of the first book.

The next and second step, though not without considerable fault, yet avoided the error of supposing that the learner was a competent critic. It consisted in treating the results of algebra as necessarily true, and as representing some relation or other, however inconsistent they might be with the suppositions from which they were deduced. So soon as it was shewn that a particular result had no existence as a quantity, it was permitted, by definition, to have an existence of another kind, into which no particular inquiry was made, because the rules under which it was found that the new symbols would give true results, did not differ from those previously applied to the old ones. A symbol, the result of operations upon symbols, either meant quantity, or nothing at all; but in the latter case it was conceived to be a certain new kind of quantity, and admitted as a subject of operation, though not one of distinct conception. Thus, $1-2$, and $a-(a+b)$, appeared under the name of negative quantities, or quantities less than nothing. These phrases, incongruous as they always were, maintained their ground, because they always produced a true result, whenever they produced any result at all which was intelligible: that is, the quantity less than nothing, in defiance of the common notion that all conceivable quantities are greater than nothing, and the square root of the negative quantity, an absurdity constructed upon an absurdity, always led to truths when they led back to arithmetic at all, or when the inconsistent suppositions destroyed each other. This ought to have been the most startling part of the whole process. That contradictions might occur, was no wonder; but that contradictions should uniformly, and without exception, lead to truth in algebra, and in no other species of mental occupation whatsoever, was a circumstance worthy the name of a mystery.

Nothing could prevail against the practical result that theorems so produced were true; and at last, when the interpretation of the abstract negative quantity shewed that a part at least of the difficulty admitted of rational solution, the remaining part, namely that of the square root of a negative quantity, was received, and its results admitted, with increased confidence.

The single algebra, when complete, leads to an unintelligible combination of symbols, $\sqrt{-1}$: not more unintelligible than was -1 when it first presented itself; for there are no degrees of absurdity in absolute contradiction of terms. The use of $\sqrt{-1}$, which leads to a variety of truths (page 41), points out that it "must have a logic" (page 41, note). I now proceed (page 92) to collect the symbols and laws of combination of algebra, or to describe *Symbolic Algebra.*

CHAPTER II.

ON SYMBOLIC ALGEBRA.

In abandoning the meanings of symbols, we also abandon those of the words which describe them. Thus *addition* is to be, for the present, a sound void of sense. It is a mode of combination represented by +; when + receives its meaning, so also will the word *addition*. It is most important that the student should bear in mind that, *with one exception*, no word nor sign of arithmetic or algebra has one atom of meaning throughout this chapter, the object of which is *symbols, and their laws of combination*, giving a *symbolic algebra* (page 92) which may hereafter become the grammar of a hundred distinct *significant algebras*. If any one were to assert that + and − might mean reward and punishment, and A, B, C, &c. might stand for virtues and vices, the reader might believe him, or contradict him, as he pleases—but not out of *this* chapter.

The one exception above noted, which has some share of meaning, is the sign = placed between two symbols, as in $A = B$. It indicates that the two symbols have the same resulting meaning, by whatever different steps attained. That A and B, if quantities, are the same amount of quantity; that if operations, they are of the same effect, &c.

The following laws are not all unconnected: but the unsymmetrical character of the exponential operation, and the want of the connecting process of + and ×, pointed out in the last chapter, renders it necessary to state them separately.

I. The *fundamental* symbols of algebra are

$$0,\ 1,\ +,\ -,\ \times,\ \div,\ (\ )^{(\ )},\ \text{and letters.}$$

In $(\ )^{(\ )}$ there is the best mode of expressing the peculiar case in which the symbol consists in position; as in A^B, in which the distinctive symbolical force of the form lies in writing B over A.

II. It is usual to call + and − *signs*, and them only: but in laying down the laws of symbolic algebra, the close connexion existing between + and − on the one hand, and × and ÷ on the other, requires that the latter should also be called *signs*. Let the former be called *term-signs*, the latter *factor-signs*. It is to insist on this connexion that I do not (for a while) introduce the more common synonymes for $A \times B$ and $A \div B$, namely AB and $\frac{A}{B}$.

III. A symbol preceded by + or − is a *term*; by × or ÷ a *factor*. In A^B, A is the *base*, B the *exponent*. When an expression consists of terms, let them be called *co-terms*; when of factors, *co-factors*.

IV. Let 0 and 1 be a *co-term* and *co-factor* of every symbol, + and × being the connecting signs of the symbol, but either + or −, either × or ÷, those of 0 and 1. As seen in

$$A = 0 + A = 1 \times A,$$
$$= A + 0 = A - 0 = A \times 1 = A \div 1,$$
$$= 0 + 1 \times A.$$

Thus 0 and 1 are a kind of initial or starting symbols, the first of terms, the second of factors.

It is seen that + and ×, placed before a symbol, do not alter it: $\times A$ is A, having reference to 1 understood, as in $1 \times A$; and $+A$ is A, having reference to 0 understood, as in $0 + A$.

V. Co-terms and co-factors which differ only in sign, are equivalent to term 0 and factor 1.

$$+A - A = 0, \quad \times A \div A = 1.$$

The more usual form of the last is $1 \times A \div A = 1$. The starting symbol is frequently used in factors, but rarely in terms. The student is well accustomed to $+A$ and $-A$, in abbreviation of $0 + A$ and $0 - A$: but not to $\times A$ and $\div A$ for $1 \times A$ and $1 \div A$. But he must use the latter a little, if he would see the complete analogy of the term and factor signs.

VI. A symbol is said to be *distributive* over terms or factors when it is the same thing whether we combine that symbol with each of the terms or factors, or whether we make it apply to the compound term or factor. Thus, looking at

$$\overbrace{ABCD}^{*} \text{ and } \overset{*}{A}\overset{*}{B}\overset{*}{C}\overset{*}{D},$$

we see the * of the first distributed in the second.

VII. Term-signs are distributive over terms, and factor-signs over factors: as in

$$+(+A-B)=+(+A)+(-B), \quad \div(\times A \div B)=\div(\times A)\div(\div B),$$

at full length $0+(0+A-B)=0+(0+A)+(0-B)$,

$$1\div(1\times A\div B) = 1\div(1\times A)\div(1\div B).$$

VIII. The term-signs of factors may belong, each one of them, to any factor of the compound, or to the compound.

$$-A\times -B = -(-A)\times B = -(-)(A\times B).$$

IX. Like term-signs in combination produce $+$; unlike, $-$. Like factor-signs in combination produce $\times$; unlike, $\div$. As in

$$+(-A)=-A,\ -(-A)=+A,\ \times(\div A)=\div A,\ \div(\div A)=\times A.$$

X. Terms and factors are convertible in order, terms with terms, factors with factors. As in

$$+A-B=-B+A, \quad \times A\div B=\div B\times A.$$

XI. Factors are distributive over the terms of any cofactor with the sign $\times$. (The corresponding law for $\div$ factors can be deduced, and is not to be set down as fundamental). As in

$$(+A)\times(+B-C)=(+A)\times(+B)+(+A)\times(-C)=+A\times B-A\times C,$$

and $\times(B-C)\div A = B\div A - C\div A$.

XII. The relations of the starting symbols 0 and 1, as exponents, are $A^0=1$, $A^1=A$.

XIII. The exponent is distributive over factors with $\times$ (the case of $\div$ is deducible). As in

$$(\times A\times B)^C = \times A^C \times B^C.$$

XIV. The operations of $\times$ and the exponential operation $(\ )^{(\ )}$, successively repeated with the same base, are reducible to the lower operations $+$ and $\times$ performed with the exponents. As in

$$A^B\times A^C = A^{B+C}, \quad (A^B)^C = A^{B\times C}.$$

Any system of symbols which obeys these rules and no others,—except they be formed by combinations of these rules—and which uses the preceding symbols and no others—except they be new symbols invented in abbreviation of combinations of these symbols—is *symbolic algebra*. Ordinary algebra contains all these symbols and all these rules, but its assigned meanings do not make *all* results significant. I now proceed to combined symbols, and to a sufficient amount of proof by instance, that one who admits these rules admits, as consequences, all the combinations of ordinary algebra.

Let $1+1$ be abbreviated into 2; $2+1$ into 3; $3+1$ into 4, and so on. Now introduce the abbreviations of $A\times B$ and $A\div B$, namely, AB and $\frac{A}{B}$.

We have then $A+A=2A$; for (IV), $A+A$ is $1\times A+1\times A$ or (XI) $(1+1)A$ or $2A$. Similarly, $A+A+A=3A$. Again, $4A\div 7$ is $\frac{4}{7}A$: for (X), $1\times 4\times A\div 7$ is $1\times 4\div 7\times A$, or $A\times 4\div 7$, or (VII), (VIII), $A\times(\times 4\div 7)$, or $\frac{4}{7}A$;

$(A-B)(C-D)$ is (XI) $(A-B)C-(A-B)D$, or, (XI) again, $AC-BC-(AD-BD)$, or (VII), $AC-BC-(+AD)-(-BD)$, or (IX), $AC-BC-AD+BD$;

$\frac{A}{B}=\frac{AC}{BC}$, for $\times A\times C\div(B\times C)$ is (VII), $\times A\times C\div B\div C$, or (X), $\times A\div B\times C\div C$, or (V), $\times A\div B$, or $\frac{A}{B}$;

$A\times 0=0$; for (V) $A\times 0$ is $A(+B-B)$, or (XI) $+AB-AB$, which (V) is 0.

From what precedes $\frac{A}{B+C}$ is $\frac{1}{\frac{B}{A}+\frac{C}{A}}$. This is an instance of the deducible part of (XI); it is $\times A\div(B+C)=\div(B\div A+C\div A)$. The complete rule XI, in all its parts, fundamental and deducible, is this:—A factor may be distributed over the terms of its cofactor, with its factor-sign or the contrary, according as the receiving cofactor is $\times$ or $\div$. Thus

$$\div A\div(B+C) \text{ is } \div(A\times B+A\times C);$$

$\frac{A}{B}\pm\frac{C}{D}$ has been shewn to be $\frac{AD}{BD}\pm\frac{CB}{BD}$, or (XI) $\frac{AD\pm BC}{BD}$.

$A^B \times A^{-B}$ is (XIV) $A^{B+(-B)}$, or (IX) A^{B-B}, or (V) A^0, or (XII) 1.

$$\text{So that } A^{-B} = \frac{1}{A^B}; \text{ and } \frac{A^B}{A^C} = A^B A^{-C} = A^{B-C};$$

$$\left(\frac{A}{B}\right)^C = (A^1 B^{-1})^C = A^C B^{-C} \text{ (XIV) } = \frac{A^C}{B^C};$$

$$A^2 \text{ is } A^{1+1}, \text{ or } A^1 A^1, \text{ or } AA; \; A^3 \text{ is } AAA, \text{ \&c.}$$

$$A^{\frac{1}{3}} \text{ gives } (A^{\frac{1}{3}})^3 = A^1 = A \text{ (XII), or } A^{\frac{1}{3}} A^{\frac{1}{3}} A^{\frac{1}{3}} = A;$$

$$-A \times -B \text{ is (VIII) } -(-) A \times B, \text{ or } +AB, \text{ or } AB;$$

$$A \times (BC) \text{ is } A \times (\times B \times C), \text{ or (VII) } A \times (\times B) \times (\times C),$$

$$\text{or (IX) } A \times B \times C.$$

In this way the student must examine narrowly a large number of fundamental operations, satisfying himself that he could produce them from the *rules alone*, independently of every notion of meaning. The question is this,—Might a machine, which could, when guided, make introductions and alterations by the preceding rules and no others, be made to turn one of the alleged equivalent combinations into the other.

It will be exceedingly convenient to reserve the small letters *a*, *b*, *c*, &c. most strictly to signify pure combinations of the unit-symbol 1, with any term or factor-signs, as $+2$, $-\frac{3}{8}$, &c.: and to use the capitals *A*, *B*, *C*, &c. for other cases. With the exception of ϵ I shall use Greek letters only for angles.

CHAPTER III.

ON AREAS AND SOLIDS.

I MAKE the first example of significant algebra to be an application of symbolic algebra to the geometry of right areas (rectangles) and right solids (rectangular* parallelepipeds), because the application is useful, and abounds in instances of the distinction between symbols which become significant under the meanings given, and those which are not significant.

However clearly a student may see that the ordinary arithmetical proofs of the propositions in the second book of Euclid are not sound, except for lines which are commensurable with one another, yet, considering that every proposition which can be proved by such arithmetical proof *must be true*† (as may be otherwise established) for all lines whatsoever, it may be suspected that the mechanism of the arithmetical proof is really the mechanism of some sound and general proof. And so it turns out, namely, that one of the significant algebras is the method of proof desired.

$+$ and $-$ are simple addition and subtraction; A, B, &c. are lines, if not otherwise specified, and it is easy to confine them to lines. Again, $\times$ in $A \times B$ makes the symbol mean the rectangle under A and B; the second $\times$ in $A \times B \times C$ makes the symbol mean the right solid under A, B, C. The symbols 0, 1, 2, &c. are as in arithmetic: thus $2AB$ is twice a rect-

* The length of this phrase is intolerable: and I am in the habit of using the following extension of the word *right*. As a right line is formed by the simplest and most direct motion of a point, so the term *right area* might be applied to that formed by the most direct motion of a right line, and *right solid* to the solid formed by the most direct motion of a right area. Accordingly, the rectangle and the rectangular parallelepiped would be the right area and right solid.

† The perfect confidence which a mathematician puts in these proofs does not arise, as he knows, from their proving that their conclusions are true, but from their proving that they can (otherwise) be proved to be true.

angle; $\times$, after a symbol derived from 1, meaning common multiplication. Exponents, save only 1, (understood, XII.) 2, and 3, need not appear. Heterogeneous terms are insignificant when put together: thus $AB + C$, the area of a rectangle added to a length, is unmeaning: as an area, the length is *nothing*. Again, $A \div B$ is merely the ratio of the two lines; all the rules become true under this meaning, joined with the others. $AB \div C$ is the other side (C being one) of the rectangle equal to the rectangle under A and B. And $ABC \div D$ is the area of the base (D being the altitude) of a right solid equal to that under A, B, C. And $ABC \div DE$ is the altitude of the same, DE being the base. And A^2 or AA is the square on A; A^3 or AAA is the cube on A.

It will be very easy now to establish that these meanings give truth to all rules which have significance: to see the following for instance.

$A(B - C) = AB - AC$, or, between the same parallels, the rectangle under the difference of two bases is equal to the difference of the rectangles under those bases.

$AB \div C = A \div C \times B$, or the remaining side (C being one) of the rectangle equal to the rectangle under A and B, is equal to the proportion of B, which is expressed by the numerical ratio of A to C.

As far as + and − are concerned, this system is that of pure arithmetic. And A^B, $ABCD$ (space not having four dimensions), are unintelligible. And we have instances of forms which are significant, while equivalent forms are insignificant. Thus $ABCD \div E$ is unintelligible; there is no solid of four dimensions. But the equivalent form of symbolic algebra, $A \div E \times BCD$ is significant: it is such proportion of the right solid BCD as A is of E. Shall we then say

$$A \div B \times BCD = ABCD \div E?$$

Shall we say, in *common* algebra,

$$1 - \frac{\theta^2}{2} + \frac{\theta^4}{2.3.4} - \ldots = \frac{\epsilon^{\theta\sqrt{-1}} + \epsilon^{-\theta\sqrt{-1}}}{2}?$$

Both questions are to be answered alike. Those who can, in common algebra, find a square root of −1, will be at no loss to find a fourth dimension of space in which ABC may become $ABCD$: or, if they cannot find it, they have but to imagine it, and call it an *impossible* dimension, subject to all the laws of the three we find possible. And just as $\sqrt{-1}$, in common algebra, gives all its *significant* combinations *true*, so would it be with any number of dimensions of space which the speculator might choose to call into *impossible* existence.

The rules having been proved true, so far as significant, all results produced by none but significant steps are pure geometry.

Thus $(A+B)^2=A^2+2AB+B^2$ is Euclid II. 4: not an arithmetical representation, but the proposition itself.

And $\frac{A^3-B^3}{A-B}=A^2+AB+B^2$, significant when $A>B$, means that the base of a right solid which equals the difference of two cubes, the difference of their sides being the altitude, is equal to the sum of the squares on the sides and their rectangle.

The student must not call this significant phase of algebra modern, though in its separated form it may be so. The superiority of the Greek geometry over Greek arithmetic, in means of expression and demonstration, caused much of the notion on which the former is founded to find its way into the latter. It is from this mixture that we get the terms *square* and *cube*, as applied to $a\times a$ and $a\times a\times a$ (numbers). Vieta, who so materially improved the symbolic language of algebra as to be rightfully considered the founder of its modern form, was so thoroughly possessed with the idea of linear, areal, and solid representation, that he would have written such an equation as $XXX+AXX+BX=C$, under the idiom

$$XXX+AXX+B\,\text{planum}\,X=C\,\text{solidum},$$

if he had used exactly our symbols. To have done otherwise, to have allowed B and C to be the same species of magnitude as X and A, would have appeared to him like asserting that two solids and an area could make a line.

I should recommend the student to consider this algebra well, and, when he meets with any circumstance of ordinary algebra in which significance is difficult to conceive or absolutely unattainable, to try if he can imagine the corresponding case of the subject of this chapter.

CHAPTER IV.

PRELIMINARY REMARKS ON DOUBLE ALGEBRA.

IF, taking the rules of symbolic algebra, we were to ask for an assignment of meaning to $(-1)^{\frac{1}{2}}$ which would make all those rules true of it, we should naturally be led to select for consideration the rule (XIV.) on which the symbolic character most depends. It is

$$(-1)^{\frac{1}{2}}(-1)^{\frac{1}{2}} = (-1)^{\frac{1}{2}+\frac{1}{2}} = (-1)^1 = -1,$$

$$\text{or} \quad -1 = \sqrt{-1} \times \{\sqrt{-1} \times 1\}.$$

Consequently, $\sqrt{-1}$ must satisfy this condition, that *twice* successively applied to $+1$ by the process of $\times$ (whatsoever that be) it has the effect of changing $+1$ into -1.

There may be many significant algebras in which this is done. But the demand made by common consent is, that our completely significant algebra shall be an *extension* of the defective system with which we commence: meaning, that so far as that system goes, significantly, it shall be a part of the new system. It would not help us, with reference to the mathematics now established, if fifty completely significant systems were produced, unless in one or more of them the same story were told as in the old algebra, so far as this last tells any story at all. We must have, if possible (and I am to show that it is possible), all that we do understand still understood in the same sense, with such enlargement of meaning as will give significance to symbols which we do not now understand. Accordingly, $+1$ and -1 are still to signify diametrically opposite units.

Let us then examine one of the usual systems of explanation, in which we have a distinct conception of two diametrically opposed directions of measurement, *and of no more*. Let it be *time*. Can we form any notion of an operation upon time ($+1$ being an hour future measured from a certain æra) which being twice repeated, shall produce an hour past, (-1)? The answer

L

seems obvious: go back an hour, and then go back an hour. But a little consideration will show that this process cannot be represented by $\sqrt{-1}$. For then $\sqrt{-1}.1$ would mean $1-1$, or 0, and $\sqrt{-1}\{\sqrt{-1}.1\}$ would be $\sqrt{-1}\{0\}$, which (page 104) must be 0. Besides, this operation, *go back an hour*, is -1; and $1-1-1$ is -1, as required. Moreover, since by the laws of algebra $\sqrt{-1}$ cannot be any positive or negative quantity, it would be absurd to say that $\sqrt{-1}.1$ could be so interpreted as to mean any time future, all which is already taken up by positive quantity, or any time past, all which is taken up by negative quantity. And we have not *any other notion of time:* there is nothing (except 0, which will not do, as seen) intermediate between time future and time past. We may then safely assert that when $+1$ means a unit of future time, -1 of past time, this algebra, significant as to all positive and negative quantity, must remain insignificant as to $\sqrt{-1}$.

Next try the simple notion of *gain and loss.* If we could imagine a commercial event, which changed £1 of otherwise certain gain into something of an intermediate character, not truly described either as gain or loss; but such that, should the event happen again, it would convert the intermediate state into £1 of certain loss—we might be prepared to hope for a significant algebra on this basis. I will not say that such a basis of significance is impossible; but only that it has never been produced, though it has been before those who think on this subject, as a suggestion, for more than forty* years. When any one shall succeed in producing such an intermediate state between gain and loss, then the symbolic algebra will become significant on a system of gain and loss.

At present, however, take what notion we may, which presents the two diametrically opposite states (page 95), we find ourselves at a loss to make a notion of anything intermediate, except in one case. *Length and direction in a plane*† offers

---

* See *Phil. Trans.*, M. Buée "Mémoire sur les Quantités Imaginaires," read in 1805: also the first edition of Dr. Peacock's Algebra, pp. 366, 367.

† I dismiss, without anything more than such allusion as will prevent my being supposed to deny them, all the bases of significance

an immediate solution of the question. We can pass from a line to its opposite, not only *along the line*, but also by *supposing the line to turn round.* The condition at the beginning of this chapter is satisfied by supposing $\sqrt{-1}\times$ to be a revolution through a right angle: for a repetition of the process turns a line through a second right angle, opposes its direction to that which it had at first, and satisfies the equation $-1=\sqrt{-1}\times\{\sqrt{-1}\times 1\}$. This observation contains the first thought which led to the inquiry into the question whether a completely significant algebra could be constructed on definitions involving, not only opposite lengths, but lengths in other directions. And hence it is frequently stated that this result is derived from assuming $\sqrt{-1}$ as the symbol of *perpendicularity.* But this statement does not give a fair representation; that $\sqrt{-1}$ represents a unit of length perpendicular to that represented by $+1$ is a *consequence,* not an *assumption:* and a consequence of assumptions of a much more simple character.

In inventing such a system, we obviously found *an* algebra on a *geometrical* basis of significance. Why this limitation? Because, except in geometry, we nowhere find the varieties of distinct conception which will afford meaning to our symbols. As before seen, we are not bound to this system. The moment any one shall afford us a distinct notion of time, or of mercantile result, intermediate between past and future, or between gain and loss, in a manner analogous to that in which a perpendicular is intermediate between the two sides of its correlative perpendicular, that moment the system of symbolic algebra is as ready to apply itself to time, or to gain and loss, as now to those geometrical ideas on which it will presently be established in significance.

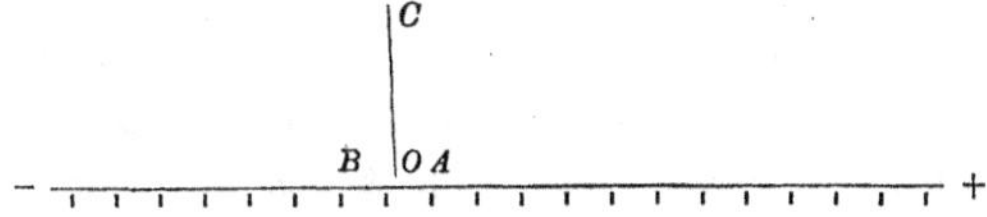

If OA stand for a pound of receipt, and OB for one of

---

which may be found in length considered in three dimensions, or on other than plane surfaces, or in lengths which are not rectilinear. The number of such bases is, I have not a doubt, quite unlimited.

expenditure, it would be perfectly easy to keep a cash account with a pair of compasses, on the line $-+$ indefinitely extended. Measure off the amount in hand at the beginning, and set off all receipts (at OA for £1) towards the right, and all out-goings towards the left, and the last point touched by the compasses would never fail to show the balance in hand. But what money does OC represent? Again, with OA to represent the first year after the Christian æra, and OB the first year before, a pair of compasses will assign its place to every stated event that ever did happen, or that we can imagine to have happened. But what event happened at OC...the Christian æra: and what adverb is proper to take the place of *before* or *after* (neither of which will do) in the blank... By such considerations we may see that we do not *restrict* ourselves to geometry, but *extend* ourselves to it: with ample means of representing all the notions we have, and introducing others for which most notions of magnitude afford us no analogues. And we may see the propriety of extending the meaning of a geometrical term, and calling time, loss and gain, &c., magnitudes of *one dimension*. But then arises the following question: Granting that we help ourselves in geometry, of what use is this algebra out of geometry, in problems which have data derived from time, or loss and gain, &c.? To put this question properly, it should be resolved into two, as follows:—

1. Suppose the problem is 'at what time after a certain epoch will an event take place which...[here describe the conditions of the problem]...' Suppose the answer to be, that the event must happen at $4+3\sqrt{-1}$ hours after the epoch: what does this mean? It means that it is really and truly *impossible** that an event should happen, under the prescribed conditions, at any imaginable moment, past or future: and that the assertion that it can happen contains the assertions that what is, is not, that a whole is no greater than its part, &c.

* The word impossible has been so misused in algebra, in the sense of *inexplicable*, that the impossible of ordinary life, which "can't be—and never, never comes to pass," requires some additional epithet to express it in an algebraical work.

As long as the meanings of symbols remain unextended,* "the essential character of imaginary expressions is to denote impossibility: and nothing can deprive them of this signification. Nothing like a geometrical construction can be applied to them; they are indications of the impossibility of any such construction, or of anything that can be exhibited to the senses."

2. Suppose that to the above problem we obtain an answer that the event takes place, say in 4 hours from the epoch, and that our solution is obtained by aid of $\sqrt{-1}$, which however disappears in the result. How are we free from the imputation of applying reasoning to contradictory terms, seeing we do not profess that, when time is the basis of significance, $\sqrt{-1}$ has any meaning. I answer that, if time continue to be our basis of significance, we are unanswerably open to that imputation: but that, if we translate the terms of our problem, that is, substitute geometrical ones, and work a geometrical answer, our whole process is intelligible; and so many units of length as our geometrical answer contains, so many units of time does the answer to the original problem contain. Algebra takes cognizance only of *units*, not of *what units they are*, whether of length or time, &c. Each of its transformations is made in one way, whatever may be the magnitudes from which the units represented by its symbols are derived. A problem given in terms of one magnitude may be solved in terms of another, provided that every condition of the problem be faithfully preserved.

---

* The quotation which follows the words in italics is from the review of M. Buée's memoir on Imaginary Quantities, in vol. XII. (1808) of the Edinburgh Review. The earlier writers on this subject were much given to supposing the explanations of $\sqrt{-1}$ to be absolute, and to be a demonstrable part of ordinary algebra: the extension of meaning, or of the field of significance, was not distinctly announced, and I imagine, indistinctly conceived. Hence, as against M. Buée, there is an amount of propriety in the reviewer's remark. But, nevertheless, it is a striking instance of the confusion between ignorance and nature, alluded to in page 98. The reviewer ought to have seen that in pure arithmetic every part of his dictum applies to *negative quantities*.

This is a point which, to the beginner, may require some illustration. Suppose, then, we have this question:—Two youths are aged 6 and 16 years; when will one be twice as old as the other? Answer, in 4 years. Now propose it thus:—Two youths have 6 and 16 apples; when will one have twice as many apples as the other? The data of the question are insufficient; there is no connexion expressed or implied between the number of apples one may get, and that which the other may get. But there *was* a connexion between their increments of age, implied in the mention of *time*, and capable of being expressed. I did not say—'Two youths are aged 6 and 16 years, *and for each year which one advances in age, the other advances a year also*, required, &c.'; because the words in italics are necessarily due to the mention of age. Now add to the second problem the condition that for each apple which either gets, the second gets one also, and we have the first problem, in which each 1 is derived from a year, faithfully rendered into another in which each 1 is derived from an apple: and the answer is, 'When each has got four apples'.

It is a true method of finding the half of ten apples, or the half of ten years, to describe an equilateral triangle upon a line of ten inches, to bisect the vertical angle (Euclid I. 11), and to show that each of the segments of the base is 5 inches. The student must take care, in applying a complete significant double algebra to questions of non-geometrical magnitude, that he does not fall into an error analogous to that of supposing the equilateral triangle to be described upon the ten years, or the ten apples.

The separation of essential from non-essential notions is a very important process to all who would think upon mathematical subjects. In the first problem we see that the answer is what it is, not because it is time of which we spoke, except so far, that between any two moments, all person's ages have received the same accession. The distance run over by persons in the same carriage would have done as well. The formation of symbolic algebra itself is a separation of the essential conditions of operation from the non-essential: the rejection of all meaning over and above the *points of meaning* on which transformations depend.

There is another instance of separation of essential notions which it will be necessary to use. In thinking of a process of arithmetic, for instance, there is the *subject-matter* of the science, and the *mode of operation:* these two things are distinct, to those who can separate them. But there may be a difficulty in doing this: is it possible, for example, that we could think of *addition* without thinking of number or magnitude, or thinking of more? This point we shall try. The subject-matter of arithmetic is *number;* its primary operation is *counting* or numeration. This counting proceeds from 0, which represents, and must represent, the state of the mind with respect to the number attained, before the counting begins. Memory (and, for high numbers, reductive modes of expression) save us from counting every time we produce number for use. Any one who had forgotten *seven* must begin as children do, first with none at all, put on *one*, put on another and say *two*, &c. until he comes to *seven.* Now let us suppose that he is to add *seven* to *three*, and that he has forgotten both seven and the total. He must proceed first by counting seven, and then by repeating the process of counting seven, with no alteration except substituting 3 in place of 0, to start from. Thus we have

In counting seven. } 0, add one, 1; add another, 2;......6; add one, 7.

In adding seven to three. } 3, add one, 4; add another, 5;......9; add one, 10.

Accordingly, a and b being two integers, the direction given in forming the arithmetical symbol $a+b$ is 'proceed from a, first formed, in the same manner as you proceed from 0 to form b.' Now if a and b *stand for* numbers, we must of course think of number in doing this. Nevertheless the *description of the operation* contains no numerical idea, except when the subject-matter is numerical. It is only 'Do with X as you did with Y to make Z,' and every book of art, on any subject whatever, abounds with this species of direction. It is seen in our symbolic algebra: for B is $0+B$; so that in $A+B$ it is seen that A only takes the place of 0.

Again, let us think of arithmetical *multiplication.* Here the separation of notion of operation from notion of subject-matter is even more easily made. What is 7 *times* 3? It is a number which has a 3 for every unit which there is in 7. The direction then is, Substitute 3 for 1 in the formation of 7. In place of $0 + 1 + 1 + 1 + 1 + 1 + 1 + 1$ write $0 + 3 + 3 + 3 + 3 + 3 + 3 + 3$. Accordingly, $a \times b$ is always the result of substituting a for 1 in the formation of b, or of proceeding with a as we proceed with unity in forming b. This is seen in the symbols: for B is $1 \times B$; and in $A \times B$, A takes the place of 1.

CHAPTER V.

SIGNIFICATION OF SYMBOLS IN DOUBLE ALGEBRA.

THIS particular mode of giving significance to symbolic algebra is named from its meanings requiring us to consider space of two dimensions (or area), whereas (page 111) all that ordinary algebra requires can be represented in space of one dimension (or length). If the name be adopted, ordinary algebra must be called *single*. I first commence with the mere description of the symbols, and then proceed to establish the rules in Chapter II.

All the symbols which in single algebra denote numbers or magnitudes, in double algebra denote *lines*, and not merely the *lengths* of lines, but their *directions*. Thus two lines of the same length, but in different directions, or two lines in the same direction, but of different lengths, *must* have different symbols. Accordingly, each symbol is meant to convey a *double* signification: it describes the length, and direction, of its line.

Two finite lines have the same direction, when they are parallel, and when they run in the same direction* on these parallels. Thus, A and B being points, AB and BA are not entitled to the same symbol: and if A, B, C, D be the points of a parallelogram in order, AB and DC have the same symbol, but not AB and CD. Thus $AB = DC$ is true: $AB = CD$ is not.

The symbol 0 has reference to one particular point, arbitrarily chosen, but steadily kept to, which may be called the *origin*. By $X = 0$, we mean that X has no length: it is the equal of a line, so to speak, which begins and ends at the origin. The line 1, is a line arbitrarily chosen as to length and direction, but steadily kept to. When 1 is drawn from the origin, the line in which it is, indefinitely extended both

* The word direction is used in two different senses. Thus north and south are different directions on a line, and the line of north and south is one direction among lines out of an infinite number.

ways, is called the *unit-line*, Afterwards, and particularly with reference to symbols of the form A^B, it will recal properties if we designate the unit-line as the *axis of length*, and the perpendicular to it as the *axis of direction*.

Since A and B are found from O by progress over certain lengths in certain directions, let us first describe the line we choose to call A, and then, proceeding from its extremity, let B be set off, commencing from the completion of A. Let the third side of the triangle, if we take the B which commences at the completion of A, or the diagonal of the parallelogram, if we take the B which commences from O, be denoted by $A+B$. Then the operative direction in page 115, is strictly applied to a different subject-matter. To form $A+B$, we put A in the place of 0 in $0+B$. And just as in arithmetic $11+7$ tells us how far we are from 0 when 7 has been counted from and after 11, so here $A+B$ is supposed to indicate how far we are from O, and in what direction, when $+B$ is joined to A. And since $(A+B-B)$ is to be $A+0$, or $0+A$, or A, annexing $-B$ must be equivalent to going over a line equal and opposite to B. And $A-B$ represents the length from O, and direction attained, by going over, first A, and then an equal and opposite to B. And $-B$, standing alone, is $0-B$, or a line equal and opposite to B from O itself.

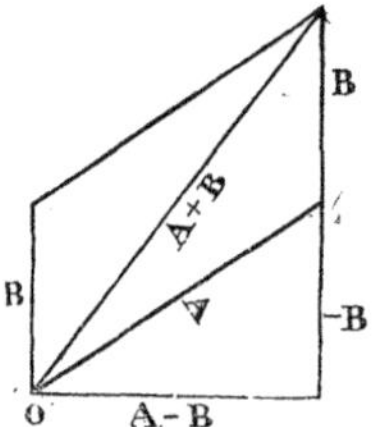

If A and B *be in the same direction*, $A+B$ and $A-B$ are as in single algebra: this will appear by following the above rules. And if we take the unit-line, it will appear that $1+1$, or 2, is *two* units of length in that line; $2+1$ *three* units of length in the same; and so on. All the symbols derived from 1, represented by small letters (p. 105) are lines in the unit-line, continued both ways: this partly appears already, and will be seen further.

It thus appears, that what we here denominate *addition* is truly not addition of *magnitude* to produce *magnitude*, but junction of *effects* to produce *joint effect*. It is the process of the seaman, when he represents himself as having only made *ten* miles (that is, on his way to port), when perhaps he has gone, on two tacks,

24 miles altogether; but his effective progress is only 10 miles. In this sense, describing two sides of a triangle, of 12 miles each, may be of no more useful effect than describing the third side of 10 miles. Nor is there, in one sense, the slightest objection to saying that 12 and 12 make 10.

Let us now consider by what process 1 (OU) becomes B. There is a change both in length and direction: the change of length is accomplished by altering OU in the ratio of OU to the length of B (or multiplying OU by the number of linear units in the length of B). The change of direction is made by turning OU through the angle made by B with OU. Now substitute A in the place of 1: multiply its length by the number of units in B, and turn it through the angle made by B with OU. This process strictly follows the direction in page 116, and if we agree that the result shall be denoted by $A \times B$, we have the following rule. The length of $A \times B$ is the *arithmetical* product of the lengths of A and B, expressed in units; and the angle of $A \times B$ with the unit-line is the *sum* of the angles of A and B.

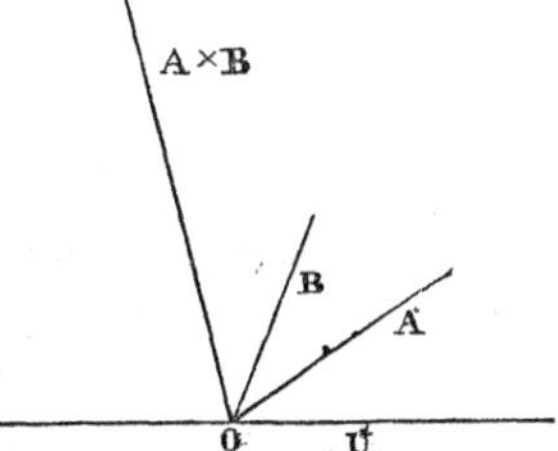

Before going further, the student must observe that we can invent a method of representing the *duplicity* of our symbols. Let letters *placed within parentheses* have their meaning in single algebra, and let (a, α) signify a line of a units of length inclined at an angle α to the unit-line. Thus 1 is (1, 0), 2 is (2, 0), &c.; -1 is $(1, \pi)$, and $(a, \alpha+2m\pi)=(a, \alpha)$. Let $A=(a, \alpha)$ and $B=(b, \beta)$, then we have

$$A \times B = (ab, \ \alpha + \beta).$$

This transformation is very easy: but addition is expressed with more difficulty. We have

$$A \pm B = \left\{\sqrt{\{a^2 + b^2 \pm 2ab \cos(\beta - \alpha)\}}, \ \tan^{-1}\frac{a \sin\alpha \pm b \sin\beta}{a \cos\alpha \pm b \cos\beta}\right\}.$$

Since $A \times B \div B$ is to be A, we have for the meaning of $A \div B$ as follows:

$$A \div B = \left(\frac{a}{b}, \ \alpha - \beta\right);$$

or the *division* of this algebra consists in dividing the length of the dividend by the length of the divisor for the number of units of length in the quotient, and subtracting the angle of the dividend from the angle of the divisor for the angle of the quotient. Observe that we need not, unless we please, use any negative number inside the parentheses: thus $(-2, a)$ is $(2, a+\pi)$ and $(2, -a)$ is $(2, 2\pi - a)$, or $(2, 4\pi - a)$, &c. Perhaps at first it will be best to avoid negative quantities within these parentheses. The following are some examples:

$$\frac{1}{B} = (1, 0) \div (b, \beta) = \left(\frac{1}{b}, 2\pi - \beta\right),$$

$$3 \times 4 = (3, 0) \times (4, 0) = (12, 0); \ \frac{3}{4} = \frac{(3, 0)}{(4, 0)} = \left(\frac{3}{4}, 0\right),$$

$$-3 \times \quad 4 = (3, \pi) \times (4, 0) = (12, \quad \pi) = -12,$$

$$-3 \times -4 = (3, \pi) \times (4, \pi) = (12, 2\pi) = (12, 0) = 12.$$

Hence it appears that in the unit-line, multiplication and division are precisely those of single algebra. But for all other directions except $(\ , 0)$ and $(\ , \pi)$, lines of the same direction have not products in that or the opposite direction.

Let AA, AAA, $AAAA$, &c., without any reference to exponents, be called the second, third, fourth, &c. *powers* of A. And let $\sqrt{A}$, $\sqrt[3]{A}$, $\sqrt[4]{A}$, &c., be lines of which the meaning is defined by $\sqrt{A} \times \sqrt{A}$, $\sqrt[3]{A} \times \sqrt[3]{A} \times \sqrt[3]{A}$, &c., being each equal to A, and they may be called the second, third, fourth, &c. *roots* of A. Then we have immediately

$$AA = (aa, 2a), \ AAA = (aaa, 3a), \ AAAA = (aaaa, 4a), \text{ \&c.},$$

$$\sqrt{A} = \left(\sqrt{a}, \frac{a}{2}\right), \quad \sqrt[3]{A} = \left(\sqrt[3]{a}, \frac{a}{3}\right), \quad \sqrt[4]{A} = \left(\sqrt[4]{a}, \frac{a}{4}\right), \text{ \&c.}$$

As explained in pages 43, 44, choice of values immediately commences, as soon as we have occasion to take a subdivision of an angle. Thus, since $a + 2m\pi$ may take the place of a, we may infer, as in the pages cited, that $\sqrt{A}$ has two directions whose angles differ by π, half a revolution; that $\sqrt[3]{A}$ has three directions, indicated by angles differing by a third of a revolution; and so on. In fact that

$$\sqrt[n]{A} \text{ is any one of } \left(\sqrt[n]{a}, \frac{a}{n} + m\frac{2\pi}{n}\right),$$

where m is any integer. Thus -1 being $(1, \pi)$ we have for

$$\sqrt{-1} \text{ either } \left(1, \frac{\pi}{2}\right) \text{ or } \left(1, \frac{3\pi}{2}\right),$$

or *the square roots of unity are units* perpendicular to the unit line.* If, to draw a distinction, we denote $(1, \frac{1}{2}\pi)$ by $\sqrt{-1}$, then $-\sqrt{-1}$ will be denoted by $(1, \frac{3}{2}\pi)$.

As yet, every symbol or combination of symbols *from the unit line*, in obeying the laws of *double* algebra, obeys also those of single algebra; the code of the latter being merely a local chapter in the code of the former. But, for symbols in general, the theorems of algebra are assertions of a much wider kind. When we say in double algebra that

$$(7 \times 7 - 2 \times 2) \div (7 - 2) = 7 + 2,$$

we repeat in substance a proposition of arithmetic, the greatest difference being that our additions and subtractions are rather carryings forwards and backwards with the compasses than numerical efforts of mind. But in establishing

$$(AA - BB) \div (A - B) = A + B,$$

we shall establish nothing less than the following geometrical theorem.

If there be two given lines inclined at given angles to a line of standard length and direction, and if to the standard and each of them a third proportional be taken, and placed at an angle with the standard double of that made by the original: and if from the end of the first line so resulting, a line be drawn equal, parallel, and opposite to the second: and if the line joining the common intersection of the standard and given lines with the last found extremity of this last line be called a *first result:* and if from the extremity of the first given line two lines be drawn equal and parallel to the second line, in the same and opposite directions: and if the lines joining the common intersection before named with the last found extremities be called *second* and *third results:* then the second result is a fourth proportional to the third result, the standard, and the

* Whatever may have been suggested by the considerations in page 109, the reader will see that double algebra is far from being *founded* on the *assumption that* $\sqrt{-1}$ *denotes perpendicularity.* If suggestion be foundation, it is more nearly founded on the separation of operation and quantity in *arithmetical* addition and multiplication.

first result, inclined to the standard at an angle equal to the excess of that of the first result over that of the third result. The student should verify some general theorems of algebra by actual drawing: this would give him practice in the meaning of the terms.

The unit line, produced both ways, might well be called the *line of single algebra;* and the positive side of it the *line of pure arithmetic.* And it readily follows that *all symbols of double algebra are capable of being expressed by symbols of single algebra, combined with* $\sqrt{-1}$: or $\sqrt{-1}$ is the only *peculiar* symbol of *double* algebra.

To show this, first observe that $a\sqrt{-1}$ is $(a, 0) \times (1, \frac{1}{2}\pi)$ or $(a, \frac{1}{2}\pi)$, or a units of length perpendicular to the unit line. Let there be a line R, and let it be projected upon the unit line

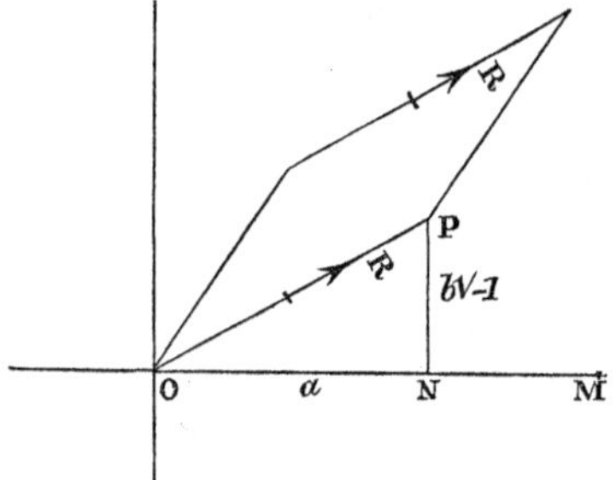

and its perpendicular into projections of a and b units of length. The first projection $(a, 0)$ is properly represented by a: but NP is $(b, \frac{1}{2}\pi)$ or $b\sqrt{-1}$: NM is b; and R is $a + b\sqrt{-1}$, by the definition of +. Thus we have a representation of any line, by means of symbols from the unit line and $\sqrt{-1}$.

Let $R = (r, \rho)$, and let the *projecting factors,* by which a line at the angle ρ is converted into its projections, be called $\cos\rho$ for the unit line, and $\sin\rho$ for the perpendicular. *Remember that we here recommence trigonometry: nothing out of my first book will be used in this second until it has been proved again as a consequence of double algebra.* We may consider $\cos\rho$ and $\sin\rho$ as by definition, the *lengths of the projections* of $(1, \rho)$. Accordingly, by similar triangles, $a = r\cos\rho$, $b = r\sin\rho$, and R or (r, ρ) is $r\cos\rho + r\sin\rho \,.\, \sqrt{-1}$. And then we have

$$(1, \rho) = \cos\rho + \sin\rho \,.\, \sqrt{-1}, \quad (r, \rho) = r(\cos\rho + \sin\rho \,.\, \sqrt{-1}).$$

I shall defer the consideration of the symbol $()^{()}$ until it has been established that all the rules in Chapter II., except XII., XIII., XIV., are necessarily true of the above symbols. Remember that the symbols in parentheses, as (a, α), are strictly those of single algebra, and can even be made those of pure arithmetic: and that those not in parentheses are *always* symbols of double algebra. Thus, at this moment, I have hardly a right to say $3 \times 4 = 12$: but in $(3, 0) \times (4, 0) = (3 \times 4, 0)$, common arithmetic gives the right to say that 3×4 *in the parentheses* is 12: so that 3×4 is $(12, 0)$ or 12 of the double algebra.

I. All the symbols have been made significant, except the exponential symbol $()^{()}$. The new symbols, $\sqrt{}$, $\sqrt[3]{}$, &c., though made significant, must be deferred till we treat of exponents.

II. III. The student may now freely use AB and $\frac{A}{B}$ for $A \times B$ and $A \div B$.

IV. In $0 + A$ we see nothing but A, or rather a case of A, which may have an infinite number of positions, and $0 + A$ is that one which begins at the origin. In $A \pm 0$ we only see injunction not to proceed from the second extremity of A in either direction. In common arithmetic, 7, for instance, written alone, might be the last 7 in 18 or any other number: but $0 + 7$ is the first 7 which is counted from 0: and 7 ± 0 is a direction not to count beyond 7, either forwards or backwards. In $1 \times A$ we have A described as the unit altered into the length of A, and made to turn through the angle of A: in $A \times 1$ or $A \div 1$ we see A described with further direction, 1 being $(1, 0)$, not to alter its length, nor its angle. In common arithmetic, 1×7 is unity altered into 7; and 7×1 or $7 \div 1$ is 7 unaltered.

V. The definitions of $-$ and $\div$ were constructed to satisfy $+ A - A = 0$, and $\times A \div A = 1$.

VI. VII. Any case of VII. may easily be shown thus. The application of $+$ or $-$ to a compound term is a direction to let the result stand, or to change it into the opposite line. Now if we apply $+$ to each of the simple terms, each of them stands, and therefore their compound stands, which is equivalent to applying the sign $+$ to the compound. But if $-$ be applied to each simple term, or if each be changed into its opposite, it will appear from common geometry that the compound is also changed into its opposite; so that the sign $-$ is applied to the compound.

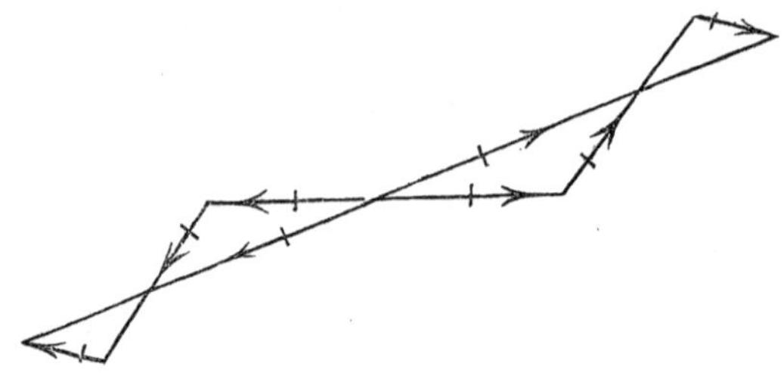

Again, in $1 \div (1 \times A \div B)$ we see that the operations are

$$1 \times A \div B = \left(\frac{a}{b}, \alpha - \beta\right), \quad (1, 0) \div (1 \times A \div B = \left\{1 \div \frac{a}{b}, 2\pi - (\alpha - \beta)\right\}$$

$$= \left(\frac{b}{a}, 2\pi + \beta - \alpha\right).$$

But in $1 \div (1 \times A) \div (1 \div B)$ we have, since

$$1 \div (1 \times A) = \left(\frac{1}{a}, 2\pi - \alpha\right), \; 1 \div B = \left(\frac{1}{b}, 2\pi - \beta\right),$$

$$\left(\frac{1}{a}, 2\pi - \alpha\right) \div \left(\frac{1}{b}, 2\pi - \beta\right), \text{ or } \left(\frac{b}{a}, \beta - \alpha\right), \text{ or } \left(\frac{b}{a}, 2\pi + \beta - \alpha\right),$$

as before.

VIII. The application of a term sign affects only the angle: nor even that, unless the sign be $-$; in which case a revolution through two right angles is produced. Now whether this alteration be made on a factor or on the whole compound, matters nothing; for whether the factor sign be $\times$ or $\div$, revolution through two right angles is of the same effect whichever way it is made.

IX. The effect of $+$ is merely permanence, that of $-$, opposition. Thus, $-(-A)$ is $+A$, for the line equal and opposite to the line equal and opposite to A must be A itself: other cases may be proved with equal ease.

Again, $\div(\div A)$ or $1 \div (1 \div A)$, A being (a, α), is $(1, 0) \underset{\div}{} \left(\frac{1}{a}, -\alpha\right)$, or $\left\{1 \div \frac{1}{a}, 0 - (-\alpha)\right\}$, or (a, α), or $1 \times A$.

X. The first part of this rule, that relating to terms, is obvious: $+A+B$ is the diagonal of a certain parallelogram, of which $+A$ and $+B$ are sides; and $+B+A$ is the same diagonal.

Hence any two consecutive terms may be made to change places; for $A + B + C - D + E = (A + B) + (+ C - D) + E$

$$= (A + B) + (- D + C) + E = A + B - D + C + E.$$

And if in any arrangement any two consecutive symbols may be made to change places, it follows that, by change after change, any one order may be converted into any other.

As to the factors, it is plain that $\times$ and $\div$ each indicates two distinct operations, either of which is capable of being performed without the other. These operations are separately of the convertible character, and their joint result is the same: for instance,

$$A \times B = (a, \alpha) \times (b, \beta) = (ab, \alpha + \beta) = (ba, \beta + \alpha) = B \times A,$$

$$\times A \div B = (a, \alpha) \div (b, \beta) = \left(\frac{a}{b}, \alpha - \beta\right) = \left(\frac{1}{b} . a, -\beta + \alpha\right) = (1 \div B) \times A.$$

XI. It may help us here, and elsewhere, to remark that there is no essential distinction between + and −, or between $\times$ and $\div$. Thus $A + B$ is $A - (- B)$, or $(a, \alpha) + (b, \beta)$ is $(a, \alpha) - (b, \pi + \beta)$. And $A \div B$ is $A \times (1 \div B)$. All cases of this rule may then be contained under

$$A (B + C) = AB + AC.$$

If any number of lines be multiplied by A, it is obvious that the products make the same angles with one another as the originals, since each angle made with the unit line is increased by α. Again, the lengths are all increased in the same proportion, their units being all multiplied by a. If then the sides and diagonal B, C, $B + C$, be all multiplied by A, we have AB, AC, $A (B + C)$, sides and diagonal of another parallelogram. Therefore $A (B + C) = AB + AC$.

With the exception of what relates to exponents, we have now a right to affirm that symbolic algebra is truly rendered significant by the preceding definitions; and that, so far, every identical equation of ordinary algebra is also an identical equation of double algebra. And further, that all ordinary or single algebra is so much of the double algebra as relates to the symbols of lines taken in the unit line or its continuation. These consequences are inevitable, unless it can be shown, first, that some indispensable rule of operation is omitted in, and cannot be deduced from, the rules in Chapter II.; and secondly, that such omitted rule, when brought forward, is found *not* to be a necessary consequence of the definitions in this chapter.

But inevitable consequences are not always easily credible: particularly when very extensive and easily deduced consequences stand upon a very small basis of definition. And it is not easily credible that the whole of trigonometry should be capable of re-establishment as a consequence of these definitions, after throwing every part of the first book away except the definitions of $\cos\theta$ and $\sin\theta$.

A close examination of all the definitions and of all the demonstrations of the symbolic rules will show that nothing of geometrical theorem is assumed except *the doctrines of parallel lines and similar triangles.* Nevertheless, what amounts to an arithmetical demonstration of Euclid I. 47, can be immediately produced.

It is seen that $(1, \theta) \times (1, -\theta) = (1, \theta - \theta) = 1$. But $(1, \theta) = \cos\theta + \sin\theta . \sqrt{-1}$ and $(1, -\theta) = \cos\theta - \sin\theta . \sqrt{-1}$, and their product is $\cos\theta\cos\theta + \sin\theta\sin\theta$, which is therefore $= 1$. Accordingly $r\cos\theta . r\cos\theta + r\sin\theta . r\sin\theta = rr$, which is the arithmetical form of I. 47. Now it is undeniable that I. 47 is proved again (without reasoning in a circle) from parallels and similar triangles in VI. 31. There must be then, in our definitions, and in the operations which are performed in

$$(\cos\theta + \sin\theta\sqrt{-1}) \times (\cos\theta - \sin\theta\sqrt{-1}),$$

something which amounts to such a deduction as is made in VI. 31. And this, it may be shown, is the fact.

Take the wider question following. From $(1, \phi) \times (1, \theta) = (1, \phi + \theta)$, we have

$$(\cos\phi + \sin\phi.\sqrt{-1})(\cos\theta + \sin\theta.\sqrt{-1}) = \cos(\phi + \theta) + \sin(\phi + \theta).\sqrt{-1}$$

$$(\cos\phi\cos\theta - \sin\phi\sin\theta) + (\sin\phi\cos\theta + \cos\phi\sin\theta).\sqrt{-1}$$
$$= \cos(\phi + \theta) + \sin(\phi + \theta).\sqrt{-1}.$$

But $a + b\sqrt{-1} = a' + b'\sqrt{-1}$ gives $a = a'$ and $b = b'$, since equal and parallel lines have equal projections. Hence we have

$$\cos(\phi + \theta) = \cos\phi\cos\theta - \sin\phi\sin\theta, \ \sin(\phi + \theta) = \sin\phi\cos\theta + \cos\phi\sin\theta.$$

Now it can be shown that the steps of the preceding multiplication are, in significance, the steps, not merely of a proof of these theorems, but of one very commonly given. Let OM be the unit line, and OM the unit, $<AOM=\theta$, $<BOM=\phi$, $<BOC=\theta$, $<COM=\phi+\theta$, and construct the obvious figure.

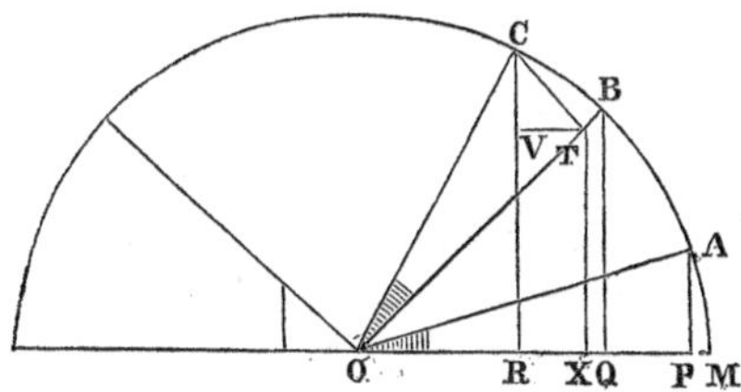

Accordingly,

$$OP=\cos\theta, \qquad OQ=\cos\phi, \qquad OR=\cos(\phi+\theta),$$

$$PA=\sin\theta.\sqrt{-1}, \quad QB=\sin\phi.\sqrt{-1}, \quad RC=\sin(\phi+\theta).\sqrt{-1},$$

$$OR=OX+XR=OX+(-RX)=OX+(-VT),$$

$$RC=RV+VC \qquad =XT+VC.$$

Now, as to lengths only,

$$OB:OT::OQ:OX \text{ or } 1:\cos\theta::\cos\phi:OX=\cos\phi\cos\theta,$$

$$OB:BQ::CT:TV \qquad VT=\sin\phi\sin\theta,$$

$$OB:BQ::OT:TX \qquad TX=\sin\phi\cos\theta,$$

$$OB:OQ::TC:VC \qquad VC=\cos\phi\sin\theta.$$

Therefore, using the geometrical designations as symbols of double algebra,

$$OX=(\cos\theta\cos\phi,\,0)\ =(\cos\phi,\,0)\ \times(\cos\theta,\,0)\ =OQ\times OP,$$

$$-VT=(\sin\phi\sin\theta,\,\pi)\ =(\sin\phi,\,\tfrac{1}{2}\pi)\times(\sin\theta,\,\tfrac{1}{2}\pi)=QB\times PA,$$

$$XT=(\sin\phi\cos\theta,\,\tfrac{1}{2}\pi)=(\sin\phi,\,\tfrac{1}{2}\pi)\times(\cos\theta,\,0)\ =QB\times OP,$$

$$VC=(\cos\phi\sin\theta,\,\tfrac{1}{2}\pi)=(\cos\phi,\,0)\ \times(\sin\theta,\,\tfrac{1}{2}\pi)=OQ\times PA,$$

$$OA\times OB=(OP+PA)\,(OQ+QB)$$

$$=OP.OQ+QB.PA+OP.QB+OQ.PA,$$

$$=OX+(-VT)+XT+VC,$$

$$=OR+RC=OC.$$

Here, first, we have formed OC, or $(1,\,\phi+\theta)$ from $OA\times OB$, or $(1,\,\theta)\times(1,\,\phi)$ with an account of the actual geometrical transformation which goes on at each step: secondly, in so doing we have been led to

$$OR = OP.OQ + QB.PA,$$

$$\text{or}\quad \cos(\phi+\theta) = \cos\phi\cos\theta + (\sin\phi\,.\sqrt{-1})\,(\sin\theta\,.\sqrt{-1})$$

$$= \cos\phi\cos\theta - \sin\phi\sin\theta,$$

$$RC = OP.QB + OQ.PA,$$

$$\text{or}\quad \sin(\phi+\theta)\,.\sqrt{-1} = (\sin\phi\,.\sqrt{-1})\cos\theta + \cos\phi(\sin\theta\,.\sqrt{-1}).$$

And by this and similar instances we may satisfy ourselves that the mechanical operations of double algebra are, when the mind takes cognizance of their significance, true proof of their results, just as is the case when they represent no more than arithmetical notions. The great difference is, that in the latter case we are much more familiar with the subject-matter, and more readily learn to make mere operation carry conviction.

CHAPTER VI.

ON THE EXPONENTIAL SYMBOL.

In proceeding to treat of exponents, it is necessary to assume the knowledge of some one system of *arithmetical logarithms.* We cannot therefore (or certainly not at first) allow the word *logarithm* to be divested of its meaning, and to pass into double algebra to receive an extended meaning. Now since our system, dealing in lines, gives results by *measurement,* the word *logometer* suggests itself as a convenient variation of the word *logarithm.* Let the logometer of A (denoted by λA) be defined as the result of some convenient operation on A which has the following property,

$$\lambda A + \lambda B = \lambda\,(AB).$$

An infinite variety of such operations may at once be given. For, since the *angle* enters in multiplication and division with the properties of a logarithm (as in—the angle of the product is the sum of the angles of the factors,—&c.), we shall find that, with respect to $(r,\ \rho)$, $M\log r + N\rho$, M and N being any fixed symbols whatever, has all the property required.

If $\qquad \lambda R = M\log r + N\rho$, we have, R' being $(r',\ \rho')$,

$$\begin{aligned}\lambda\,\{RR' \text{ or } (rr',\ \rho+\rho')\} &= M\log rr' + N(\rho+\rho')\\ &= M\log r + N\rho + M\log r' + N\rho'\\ &= \lambda R + \lambda R'.\end{aligned}$$

Now as we wish to preserve the single algebra intact as to all unit-line symbols, we must make $M = 1$; for otherwise $\lambda(r,\ 0)$ would not be $\log r$. As to N, the most convenient assumption is the form $k\sqrt{-1}$, which would give $\lambda R = \log r + k\rho.\sqrt{-1}$. But it does not limit us if we make $k = 1$: for neither the base of the logarithms nor the mode of choosing an angular unit is yet settled (in *this* book), and the power of changing the value of k is supplied by that of changing the unit in which the angle is expressed. Our definition of λR is now contained in

$$\lambda R = \log r + \rho\sqrt{-1},$$

or the logometer of any line has the logarithm of the length for its projection on the unit line, and the angle (meaning a line of as many linear units as the angle has of angular units) for its projection on the perpendicular. And this is the connexion of the two axes with length and direction from which the terms suggested in page 118 are derived. Thus we have, it appears, a species of logarithm to R on each axis; or a symbol which is augmented by addition in multiplication, &c.

In this symbol, λR, occurs, for the first time, a choice of meanings; and that choice is unlimited. For ρ we may write $\rho \pm 2m\pi$ without altering the meaning of R: but for each value of m we have a distinct logometer to R. And all the logometers of R are diagonals of rectangles standing on one base, $\log r$, with altitudes ρ, $\rho + 2\pi$, $\rho + 4\pi$, &c., $\rho - 2\pi$, $\rho - 4\pi$, &c.

But though every line have an infinite number of logometers, yet every logometer has only one primitive line. For if $a + b\sqrt{-1}$ be a logometer, its primitive can have no length except the number whose logarithm is a, and can be in no direction except that indicated by the angle b. Consequently, if two primitives be equal, we can only say that *any* logometer of the first is *one* of the logometers of the second: but if two logometers be equal, we can assert that *the* primitive of the first is equal to *the* primitive of the second.

Now take the following as the definition of the symbol A^B. Let its logometer be $B\lambda A$: that is, let it mean the line whose logometer is $B\lambda A$. If we use, for a little while, the inverted letter $\curlyvee$, so that $\curlyvee A$ shall signify the line whose logometer is A, we may state that we use A^B as an abbreviation of $\curlyvee(B\lambda A)$.

Let ε be the base of the arithmetical logarithms used in the unit line: and remember that ε is strictly to denote a line at an angle 0 to the unit line; it is $(\varepsilon, 0)$, and $(\varepsilon, 2\pi)$, or $(\varepsilon, 4\pi)$ is, for a base of logarithms, or in connexion with logometers, to be distinguished from $(\varepsilon, 0)$. Also, observe that in like manner as we have abandoned (till we recover it) the mode of measuring angles, so we do not yet say that ε is the peculiar base of the Naperian logarithms: let it be any which it can be, until we show cause for preferring one base to another. We have then

$$\lambda R = \log r + \rho\sqrt{-1} = \left\{\sqrt{(\log^2 r + \rho^2)}, \quad \tan^{-1}\frac{\rho}{\log r}\right\},$$

$$\curlyvee R = (\text{number whose log. is } r\cos\rho, \quad r\sin\rho),$$

$$\lambda\,(a + b\sqrt{-1}) = \tfrac{1}{2}\log\,(a^2 + b^2) + \tan^{-1}\frac{b}{a}.\sqrt{-1},$$

$$\curlyvee\,(a + b\sqrt{-1}) = (\text{number whose logarithm is } a,\ b).$$

Having shown that the fundamental formulæ of trigonometry are deducible from the double algebra, I now use the first book of this treatise, in every point except specifying the mode of taking an angular unit.

The following are the proofs that, under the above definition of A^{B}, the laws of symbolic algebra are true.

XII. In A^0, we are to see $\curlyvee(0 \times \lambda A)$, or $\curlyvee(0 + 0\sqrt{-1})$, or $(1, 0)$, or 1. In A', we have $\curlyvee(1 \times \lambda A)$, or $\curlyvee\lambda A$, or A.

XIII. We prove two symbols identical in meaning, if we prove any one logometer of the first equal to any one of the second. Now, by definition, the logometer of $(AB)^{\mathrm{C}}$ is $C\lambda\,(AB)$, or $C(\lambda A + \lambda B)$, or $C\lambda A + C\lambda B$. But the logometer of $A^{\mathrm{C}}B^{\mathrm{C}}$ is $\lambda A^{\mathrm{C}} + \lambda B^{\mathrm{C}}$, or $C\lambda A + C\lambda B$. Therefore $(AB)^{\mathrm{C}} = A^{\mathrm{C}}B^{\mathrm{C}}$, provided that we use the same logometers of A on both sides, and the same of B: that is, the same cases of $\alpha + 2m\pi$ and $\beta + 2m\pi$ on both sides.

XIV. Since $\lambda(A^{\mathrm{B}}A^{\mathrm{C}})$ is $\lambda A^{\mathrm{B}} + \lambda A^{\mathrm{C}}$, or $B\lambda A + C\lambda A$, or $(B + C)\lambda A$, which is a logometer of $A^{\mathrm{B+C}}$, it follows that

$$A^{\mathrm{B}}A^{\mathrm{C}} = A^{\mathrm{B+C}},$$

if we use the same logometer of A throughout.

Again, $\lambda\{(A^{\mathrm{B}})^{\mathrm{C}}\} = C\lambda A^{\mathrm{B}} = CB\lambda A$, which is a logometer of A^{BC}.

Therefore $$(A^{\mathrm{B}})^{\mathrm{C}} = A^{\mathrm{B}},$$

provided the same logometer of A be used on both sides.

Next it is to be shown that *when the exponent is a symbol of the unit line*, as in A^m, the above definitions of the exponent agree with those of ordinary algebra. This is in fact, contained above: for A^2 is A^{1+1}, or A^1A^1, or AA; $A^{\frac{1}{3}}$ is such that

$$A^{\frac{1}{3}}A^{\frac{1}{3}}A^{\frac{1}{3}} = A^{\frac{3}{3}} = A\,;$$

whence $A^{\frac{1}{3}}$ is $\sqrt[3]{A}$; and so on. But, once for all,

$$A^m = \curlyvee(m\lambda A) = \curlyvee(m \log a + ma\sqrt{-1}) = (a^m,\ ma),$$

from which all the cases can be deduced.

For $\lambda(\epsilon, 0)$, we have $1 + 0\sqrt{-1}$ or 1. Hence for ϵ^A we have $\curlyvee(A\lambda\epsilon)$ or $\curlyvee A$. That is, ϵ^A must be our future way of expressing $\curlyvee A$; and we have, as in ordinary algebra, $A = \epsilon^{\lambda A}$. Again, $(1, \theta)$ has for its logometer $0 + \theta\sqrt{-1}$; therefore $(1, \theta)$ is $\epsilon^{\theta\sqrt{-1}}$. But this is $\cos\theta + \sin\theta.\sqrt{-1}$: therefore

$$\epsilon^{\theta\sqrt{-1}} = \cos\theta + \sin\theta.\sqrt{-1}.$$

Here again occurs the difficulty of page 126. We get this fundamental equation on terms so cheap, that we suspect its goodness. And moreover, it cannot be always true, while ϵ and the angular unit are both unnamed. The second side does not depend for its numerical value upon what number θ is, but only upon what angle it represents. The first side is dependent for its numerical value upon those of ϵ and θ. If, for instance, we choose to halve the angular unit, so that the angle now containing θ units contains 2θ units, the second side is unaltered. But

$$\cos 2\theta + \sin 2\theta.\sqrt{-1} \text{ is } \epsilon^{2\theta\sqrt{-1}},$$

which is not $\epsilon^{\theta\sqrt{-1}}$. Nevertheless it will be easy both to establish that some such equation must exist, and that a connexion exists between the base to be taken for the logarithmic system and the unit of angular measure.

Having established all the fundamental rules, we may by the process in page 205 of the Algebra, interpreting the symbols as in the double system, show that any function which possesses the property $fA \times fB = f(A + B)$ must be of the form C^A, where C is independent of A: and $\cos\theta + \sin\theta.\sqrt{-1}$ is such a function of θ. Accordingly we must have

$$C^\theta = \cos\theta + \sin\theta.\sqrt{-1}.$$

This result only differs from the former in that C, which is quite arbitrary, takes the place of $\epsilon^{\sqrt{-1}}$, which is wholly undetermined. Return to the first, and observe that it gives

$$\varepsilon^{\sqrt{-1}} = \cos 1 + \sin 1.\sqrt{-1},$$

which is the relation that must exist between the angular unit and the base of the logarithms. If we were to make no appeal to common algebra, we should proceed with this equation to define the above relation, and in process of time we should arrive at the result that *if* the method of angular measurement be arcual, the base of the logarithms must be Naperian; that is, that if angle 1 have an arc equal to the radius, ε must be

$$1+1+\frac{1}{2}+\frac{1}{2.3}+\ldots.$$

But as it may be unsatisfactory to leave such a point behind us, I will establish it on the following basis: the binomial theorem with a positive integer exponent, and the theorems that $\log(1+x)\div x$ and $\tan x \div x$ both have the limit unity when x is diminished without limit; with, of course, the explained symbols of double algebra. And in assuming $\tan x \div x$ to have 1 for its limit, we assume the arcual unit.

Let us consider $\left(1+\frac{\sqrt{-1}}{R}\right)^{R}$. Its logometer is $R\lambda\left(1+\frac{\sqrt{-1}}{R}\right)$. Now

$$\sqrt{-1}\div R \text{ or } \left(1,\ \frac{\pi}{2}\right)\div(r,\ \rho) \text{ is } \left(\frac{1}{r},\ \frac{\pi}{2}-\rho\right), \text{ say } \left(k,\ \frac{\pi}{2}-\rho\right)$$

(page 119) $(1,0)+\left(k,\ \frac{\pi}{2}-\rho\right)=\left\{\sqrt{(1+2k\sin\rho+k^2)},\ \tan^{-1}\left(\frac{k\cos\rho}{1+k\sin\rho}\right)\right\}$

$$\lambda\left(1+\frac{\sqrt{-1}}{R}\right)=\tfrac{1}{2}\log(1+2k\sin\rho+k^2)+\tan^{-1}\frac{k\cos\rho}{1+k\sin\rho}.\sqrt{-1}.$$

And, R being $(\cos\rho+\sin\rho.\sqrt{-1})\div k$, we have

$$R\lambda\left(1+\frac{\sqrt{-1}}{R}\right)=P+Q\sqrt{-1}, \text{ where}$$

$$P=\frac{\cos\rho}{2k}\log(1+2k\sin\rho+k^2)-\frac{\sin\rho}{k}\tan^{-1}\frac{k\cos\rho}{1+k\sin\rho},$$

$$Q=\frac{\sin\rho}{2k}\log(1+2k\sin\rho+k^2)+\frac{\cos\rho}{k}\tan^{-1}\frac{k\cos\rho}{1+k\sin\rho}.$$

Now let the length of R increase without limit, or let k diminish without limit. Then we have

$$\frac{\log(1+2k\sin\rho+k^2)}{k}=\frac{\log(1+2k\sin\rho+k^2)}{2k\sin\rho+k^2}.(2\sin\rho+k);\text{ limit, } 2\sin\rho:$$

and, taking that angle which diminishes without limit with its tangent, we have

$$\frac{1}{k}\tan^{-1}\frac{k\cos\rho}{1+k\sin\rho}=\tan^{-1}\frac{k\cos\rho}{1+k\sin\rho}\div\frac{k\cos\rho}{1+k\sin\rho}\times\frac{\cos\rho}{1+k\sin\rho},$$

and the limit is $\cos\rho$. Hence P has for its limit

$$\tfrac{1}{2}\cos\rho.2\sin\rho-\sin\rho\cos\rho,\ \text{ or } 0:$$

$$Q \text{ has } \tfrac{1}{2}\sin\rho.2\sin\rho+\cos\rho.\cos\rho,\ \text{ or } 1;$$

and $$P+Q\sqrt{-1} \text{ has } \sqrt{-1}.$$

If then the length of R increase without limit, $(1+\sqrt{-1}\div R)^{R}$ has for its limit $\curlyvee(0+1\sqrt{-1})$ or $(1, 1)$, provided that the logometer used have an angle between $-\pi$ and $+\pi$, and that the logarithms used be of the system which gives $\log(1+x)\div x$ the limit unity. Let $R=n\sqrt{-1}$, n being integer; then

$$\left(1+\frac{1}{n}\right)^{n\sqrt{-1}} \text{ or } \left\{\left(1+\frac{1}{n}\right)^{n}\right\}^{\sqrt{-1}} \text{ has the limit } (1,1):$$

but, as n increases without limit, $(1+1\div n)^{n}$ approaches the limit $1+1+\frac{1}{2}+\dots$ (Algebra, page 225). Consequently we have

$$\left(1+1+\frac{1}{2}+\frac{1}{2.3}+\dots\right)^{\sqrt{-1}} \doteq \cos 1+\sin 1.\sqrt{-1},$$

or ϵ has the value used for that letter in single algebra.

We have now a completely significant system of algebra, and the whole contents of Book I. Chapter V. are established by demonstration, if that chapter be now inserted here.

The symbol λR is the first in which multiplicity of meaning occurs; a property which it communicates to R^{S}. All the meanings of this last symbol, the distinction between the cases in which their number is infinite and those in which it is finite, &c., will be best seen by reducing R^{S} to another form. Let

$$R=a+b\sqrt{-1},\quad S=p+q\sqrt{-1},$$

$$R^{S}=(\varepsilon^{\lambda R})^{S}=\varepsilon^{S\lambda R}=\varepsilon^{\left\{\frac{1}{2}\log(a^2+b^2)+\tan^{-1}\frac{b}{a}.\sqrt{-1}\right\}\{p+q\sqrt{-1}\}}$$

$$=\varepsilon^{\frac{p}{2}\log(a^2+b^2)-q\tan^{-1}\frac{b}{a}+\left\{\frac{q}{2}\log(a^2+b^2)+p\tan^{-1}\frac{b}{a}\right\}\sqrt{-1}}$$

$$=(a^2+b^2)^{\frac{p}{2}}\,\varepsilon^{-q\tan^{-1}\frac{b}{a}}\,\varepsilon^{\left\{\frac{q}{2}\log(a^2+b^2)+p\tan^{-1}\frac{b}{a}\right\}\sqrt{-1}}$$

$$=\left\{(a^2+b^2)^{\frac{p}{2}}\,\varepsilon^{-q\tan^{-1}\frac{b}{a}},\ \frac{q}{2}\log(a^2+b^2)+p\tan^{-1}\frac{b}{a}\right\};$$

since $r\varepsilon^{\theta\sqrt{-1}}$ is $r\cos\theta + r\sin\theta.\sqrt{-1}$ or (r, θ). Here (page 46) $\tan^{-1}(b \div a)$ may be any angle with $b \div a$ for its tangent, in which the cosine and sine have the sign of a and b; and $(a^2 + b^2)^{\frac{p}{2}}$ is taken with a positive sign.

First, it appears that when q is not $= 0$, that is, when the exponent is *not* a symbol of single algebra, the number of values is absolutely unlimited. But even in this case, when p is a rational fraction, the number of *directions* is no more than one for each unit of the denominator: and when p is an integer, there is only one direction (pages 43, 44).

Next, when $q = 0$, we have

$$(a + b\sqrt{-1})^p = \left\{(a^2 + b^2)^{\frac{p}{2}}, \quad p\tan^{-1}\frac{b}{a}\right\},$$

which has only one length, and as many directions as there are units in the denominator of p. If p be incommensurable, the number of variations of direction is infinite. The case of $b = 0$ is discussed in pages 45, 46.

The effect of the term $q\sqrt{-1}$ in the exponent, is the addition of $\frac{1}{2}q\log(a^2 + b^2)$ to the angle, and the subtraction of $q\tan^{-1}(b \div a)$ from the logarithm of the length.

The student should exercise himself in the reduction of different forms of R^S to significance, first, by the complete process, next by the formular result. For instance, $\sqrt{-1}^{\sqrt{-1}}$. Here $\sqrt{-1}$ is any case of $(1, 2m\pi + \frac{1}{2}\pi)$, and its logometers are contained in

$$(2m\pi + \tfrac{1}{2}\pi)\sqrt{-1}, \quad \text{or } (2m\pi + \tfrac{1}{2}\pi, \quad 2n\pi + \tfrac{1}{2}\pi). \quad \text{But}$$

$$(2m\pi + \tfrac{1}{2}\pi, \quad 2n\pi + \tfrac{1}{2}\pi) \times (1, \quad 2k\pi + \tfrac{1}{2}\pi) = (2m\pi + \tfrac{1}{2}\pi, \quad 2n\pi + \pi),$$

for it is not worth while to distinguish $2n\pi$ and $2n\pi + 2k\pi$, n and k being any integers we please. This last is the logometer of the result required; therefore

$$\sqrt{-1}^{\sqrt{-1}} = \left\{\varepsilon^{(2m+\frac{1}{2})\pi\cos\pi}, \ (2m + \tfrac{1}{2})\pi\sin\pi\right\} = \left\{\varepsilon^{-(2m+\frac{1}{2})\pi}, \ 0\right\} = \varepsilon^{-(2m+\frac{1}{2})\pi}.$$

Otherwise $\sqrt{-1} = 0 + 1\sqrt{-1} = \varepsilon^{(2m+\frac{1}{2})\pi\sqrt{-1}}$, $\sqrt{-1}^{\sqrt{-1}} = \varepsilon^{-(2m+\frac{1}{2})\pi}$,

or $\sqrt{-1}^{\sqrt{-1}}$ is any power of $\varepsilon^{-\frac{1}{2}\pi}$, whose exponent is in the series $\ldots -3, 1, 5, 9$, &c.

The following fallacy has before now been seriously proposed as an argument against the introduction of *imaginary* quantities into ordinary algebra. Since $1^a = 1$, $1^{\sqrt{-1}} = 1$: but $1 = \varepsilon^{2\pi\sqrt{-1}}$, therefore $(\varepsilon^{2\pi\sqrt{-1}})^{\sqrt{-1}} = 1$ or $\varepsilon^{-2\pi} = 1$, which is absurd. If we try $1^{\sqrt{-1}}$, we have $\lambda 1 = 2m\pi\sqrt{-1}$, $\sqrt{-1}\,\lambda 1 = -2m\pi$, which is the logometer of $\varepsilon^{-2m\pi}$. Accordingly, we admit the equation $1^{\sqrt{-1}} = \varepsilon^{-2m\pi}$, meaning that if m be any integer, positive or negative, $\varepsilon^{-2m\pi}$ is *one* of the values of $1^{\sqrt{-1}}$. And if $m = 0$, this is 1. But this last 1 is not $\varepsilon^{2\pi\sqrt{-1}}$: the first 1 is (1, 0), the second is (1, 2π). How these should give different logometrical results, double algebra makes manifest enough. The logometric operation makes differences of *form and value both* out of *differences of form without difference of value.*

In the Rules XIII. and XIV. it is demanded that the same logometers of each symbol shall be used throughout; otherwise the relations are not true. Does the neglect of any analogous regulation lead to errors in single algebra? To try this, let us see if error may be produced. First take $A^B . A^C = A^{B+C}$, and observe that $\lambda P = \lambda Q + R$ means $P = Q\varepsilon^R$. Use a particular logometer in A^B, call it $\lambda_1 A$, and another, $\lambda_1 A + 2m\pi\sqrt{-1}$, in A^C. The logometer of $A^B A^C$, thus taken, is

$$B\lambda_1 A + C(\lambda_1 A + 2m\pi\sqrt{-1}), \text{ or } \lambda_1 A^{B+C} + 2m\pi C\sqrt{-1}.$$

Hence $$A^B . A^C = A^{B+C}\,\varepsilon^{2m\pi C\sqrt{-1}},$$

in which A^B and A^{B+C} are formed from the same logometer. A very simple instance of the truth of *this* equation will show that beginners may commit a mistake in ordinary algebra. Let $B = C = \frac{1}{2}$, $m = 1$. Then we have $A^{\frac{1}{2}}A^{\frac{1}{2}} = A^1\epsilon^{\pi\sqrt{-1}} = -A$. But it ought to be $+A$. This beginner's mistake is like the following: $\sqrt{4} = +2$, $\sqrt{4} = -2$, therefore $\sqrt{4}\,.\,\sqrt{4} = -4$, or $+4 = -4$. The two different forms of $A^{\frac{1}{2}}$ are formed from different logometers.

Unity, when exhibited in the form $\varepsilon^{2m\pi\sqrt{-1}}$, is formed from the logometer $0 + 2m\pi\sqrt{-1}$: and a, exhibited as $a\epsilon^{2m\pi\sqrt{-1}}$, is formed from the logometer $\log a + 2m\pi\sqrt{-1}$. If we consider as *primary* that form of a symbol which takes its angle from the *first positive* revolution, or from 0 inclusive to 2π exclusive, and if $\lambda_0 A$ denote the primary logometer thence formed, and if $\lambda_m A$ denote the logometer $\log a + (\alpha + 2m\pi)\sqrt{-1}$, and $A_m{}^B$ the value of A^{B} formed from it, we have the following equations:

$$A_m{}^{\mathrm{B}} = A_0{}^{\mathrm{B}}\, \varepsilon^{2m\pi\mathrm{B}\sqrt{-1}}, \qquad A_m{}^{\mathrm{B}} = A_n{}^{\mathrm{B}}\varepsilon^{2(m-n)\pi\mathrm{B}\sqrt{-1}},$$

$$A_m{}^{\mathrm{B}} A_n{}^{\mathrm{B}} = A_p{}^{\mathrm{B}}\varepsilon^{2(m+n-p)\pi\mathrm{B}\sqrt{-1}}, \quad (A_m B_n)^{\mathrm{C}} = A_k{}^{\mathrm{C}} B_l{}^{\mathrm{C}}\, \varepsilon^{2(m+n-k-l)\pi\mathrm{C}\sqrt{-1}},$$

$$(A_m{}^{\mathrm{B}})_n{}^{\mathrm{C}} = A_k{}^{\mathrm{BC}}\varepsilon^{\{2(m-k)\mathrm{BC}+2n\mathrm{C}\}\pi\sqrt{-1}}.$$

The following equation, $\dfrac{\log(-1)}{\sqrt{-1}} = \pi$, very often found in books of algebra, merely means, when brought to significance by adoption into double algebra, that $\pi\sqrt{-1}$ is one meaning of $\lambda(-1)$.

In former days, it was not uncommon to object to the equation $\sqrt{-1}\,.\sqrt{-1} = -1$, on the ground that it should be $\sqrt{(-1 \times -1)}$ or $\sqrt{1}$ or ± 1. But it was hardly seen that, on this mode of reasoning, $\sqrt{a}.\sqrt{a}$ is $\sqrt{a^2}$ or $\pm a$, in all cases. And moreover this last *is* true, if $\sqrt{a}$ be indefinite. For then it has two values; and if in $\sqrt{a}\,.\sqrt{a}$, we are not bound to use the same value in both the first and second factor, then $\sqrt{a}\,.\sqrt{a}$ *is* $\pm a$, $+a$ or $-a$, according as $\sqrt{a}$ and $\sqrt{a}$ represent the same or different square roots. The two square roots of a are constructed on different logometers; one on $\log a + 2m\pi\sqrt{-1}$, the other on

$$\log a + (2m+1)\,\pi\sqrt{-1}.$$

CHAPTER VII.

MISCELLANEOUS REMARKS AND APPLICATIONS.

THE theory of logarithms admits, and even requires, an extension above what has been given to it. The *logometer* of the last chapter answers to the ordinary Naperian logarithm of algebra; we are now to examine what answers to the logarithm *to any base.*

It will, at a future time, when a significant algebra is made the basis of elementary instruction, be a question whether the symbol should not indicate the *amount of revolution* of a line as well as its *length* and *direction*: whether, for instance, (a, α) and $(a, \alpha + 2\pi)$ should not be distinguished by some difference of symbol. But even at present, in all that relates to logometers, it will be convenient to adopt this distinction. Accordingly R may denote (r, ρ), ρ lying between 0 and 2π; while R_m may denote $(r, \rho + 2m\pi)$.

Let B or (b, β) be the base; it is required to find the logometer of X or (x, ξ) to this base, defined by the equation $B^{\lambda_B X} = X$. The logometer of the last chapter has $(\varepsilon, 0)$ for its base. Denoting $\lambda_\varepsilon X$ simply by λX, we have $\lambda_B X . \lambda B = \lambda X$, or

$$\lambda_B X = \frac{\lambda X}{\lambda B} \frac{\log x + \xi\sqrt{-1}}{\log b + \beta\sqrt{-1}}.$$

The extension in page 48, supposes B to be $(\varepsilon, 2m\pi)$, and X to be a symbol of the unit line, or $\xi = 2n\pi$ or $(2n+1)\pi$. When

$$\log x : \log b :: \xi : \beta,$$

we have $\qquad \lambda_B X = \log x : \log b = \log(\text{base } b)\ x.$

That $\phi(a + b\sqrt{-1})$ must take the form $p + q\sqrt{-1}$, a proposition collected by a laborious induction in incomplete algebra, is now no more than was, in that algebra, the assertion that a real function of a real quantity is a real quantity. For every combination of symbols can be explained, and everything explicable is a line of definite length and direction, and every such line can be represented by $p + q\sqrt{-1}$. Nor is it more difficult to prove that if ϕx be a real, or, as we should now call it, *unit line*, function of x, $\phi(x - y\sqrt{-1})$ must be $p - q\sqrt{-1}$. For the same operations, performed on the same lines, will produce the same resulting lines, by whatever symbols they are denoted. Change the positive and negative directions on the axis of direction, and also the positive and negative directions of revolution. All the symbols of the unit line still represent what they did before; but the lines which were $a + b\sqrt{-1}$ and $p + q\sqrt{-1}$, are now $a - b\sqrt{-1}$, and $p - q\sqrt{-1}$. Therefore, the same operations on the same lines producing the same result, we have $\phi(a - b\sqrt{-1}) = p - q\sqrt{-1}$: but if the function ϕ should contain other double symbols, as $a' + b'\sqrt{-1}$, &c. and if

$$\phi(a + b\sqrt{-1},\ a' + b'\sqrt{-1},\ \&c.) = p + q\sqrt{-1},$$

then

$$\phi(a - b\sqrt{-1},\ a' - b'\sqrt{-1},\ \&c.) = p - q\sqrt{-1}.$$

It would seem as if there is still left one source of inexplicable result: what is the *angle* $\alpha + \beta\sqrt{-1}$? what is the *length* $a + b\sqrt{-1}$? or what is meant by the symbol $(a + b\sqrt{-1},\ \alpha + \beta\sqrt{-1})$? There is nothing here except such a confusion of symbols as arises in arithmetic when $7 + 4 - 5$, for instance, is by mere inadvertence of operation presented as $7 + (4 - 5)$. We have

$$a + b\sqrt{-1} = \epsilon^{\frac{1}{2}\log(a^2+b^2) + \tan^{-1}\frac{b}{a}\cdot\sqrt{-1}},$$

$$(a + b\sqrt{-1})\,\epsilon^{(\alpha+\beta\sqrt{-1})\sqrt{-1}} = \epsilon^{\frac{1}{2}\log(a^2+b^2)-\beta}\,\epsilon^{(\tan^{-1}\frac{b}{a}+\alpha)\sqrt{-1}}$$

$$= \{\sqrt{(a^2 + b^2)}\,\epsilon^{-\beta},\ \tan^{-1}\frac{b}{a} + \alpha),$$

a line of intelligible length and direction. The suppositions which, assuming (r, ρ) as the solution of a problem, end with $r = a + b\sqrt{-1}$, $\rho = \alpha + \beta\sqrt{-1}$, are analogous to those which in ordinary algebra introduce the impossible subtraction into the process of solution, when it is not necessarily produced in the answer.

If the assumption of a length a, should lead to $\alpha + \beta\sqrt{-1}$ as the requisite angle, it means that the length a will not do, but that $a\epsilon^{-\beta}$ will do, with the angle α.

If we extend the definitions of $\cos\theta$ and $\sin\theta$ so as to derive them from the equations

$$\cos\theta = \frac{\epsilon^{\theta\sqrt{-1}} + \epsilon^{-\theta\sqrt{-1}}}{2},\quad \sin\theta = \frac{\epsilon^{\theta\sqrt{-1}} - \epsilon^{-\theta\sqrt{-1}}}{2\sqrt{-1}},$$

we have, when θ is a unit-line symbol, their meanings unaltered; and when θ is not, still an intelligible signification. Thus

$$\cos(\theta\sqrt{-1}) = \frac{\epsilon^{\theta}+\epsilon^{-\theta}}{2}, \quad \sin(\theta\sqrt{-1}) = \frac{\epsilon^{\theta}-\epsilon^{-\theta}}{2}\sqrt{-1}, \text{ \&c.}$$

Similarly $\sin^{-1}(x\sqrt{-1})$, &c. can be interpreted.

The notion of continuity generally derived from ordinary algebra is corrected in the double system. If a unit symbol gradually change from positive to negative, passing through 0, there is at the moment of passing through 0, an instantaneous accession of π to the angle, and $\pi\sqrt{-1}$ to the logometer. Accordingly, the square roots are at once advanced by $\frac{1}{2}\pi$, the cube roots by $\frac{1}{3}\pi$; and so on. But we are apt to think *only* of length, which, in the case in question, does change continuously. The only perfectly continuous way of passing from $x = +a$ to $x = -a$, is by supposing θ to change from 0 to π, or from 0 to $-\pi$, in the formula $x = a(\cos\theta + \sin\theta\sqrt{-1})$: and the corresponding continuous passage from $x = a$ to $x = b$ is obtained by the same change made in

$$x = \tfrac{1}{2}(a+b) + \tfrac{1}{2}(a-b)\cos\theta + \tfrac{1}{2}(a-b)\sin\theta\,.\sqrt{-1}.$$

In this change all the roots also change continuously. In many parts of the integral calculus, results which are inexplicable on the supposition of change of length, are at least intelligible on the supposition of revolution of length, though the connexion of the two is not yet elucidated.

When the data of a problem are those of the significant system, any one of the problems which are really impossible while the terms are those of ordinary algebra, becomes possible as soon as the terms are allowed the extension of double algebra. For instance, it is required to divide $2a$ into two *parts* with the product b. The parts are $a+\sqrt{(a^2-b)}$ and $a-\sqrt{(a^2-b)}$. If a and b be numbers, the problem is arithmetically soluble if a^2-b be positive: that is, if a and b be unit-line symbols, the *parts* required in the problem are also unit-line symbols. But if a^2-b be negative, the parts are

$$a \pm \sqrt{(b-a^2)}\,.\sqrt{-1}, \quad \text{or} \quad \left\{\sqrt{b},\ \pm\tan^{-1}\frac{\sqrt{(b-a^2)}}{a}\right\},$$

the product of these is $(b, 0)$ or b, and their sum is a, for the parts are the sides of an equilateral parallelogram, of which $(a, 0)$ is the diagonal. But the *parts* are not now entitled to that name, arithmetically speaking: they are *components*, but under a law of composition which is not merely addition of magnitude.

The following theorem, given by M. Cauchy for the determination of the number of imaginary roots of an equation, and for the proof that every equation has as many roots as dimensions, can be established with clearness by the use of double algebraical meanings.

Let x and y be the projections of z in (z, ζ), or the coordinates of the extremity of Z. Let $\phi Z = AZ^n + BZ^{n-1} + \ldots$ an integral function of Z: A, B, &c. being symbols each of which has only one value, or at least, of which only one value is to be here employed. Write $x + y\sqrt{-1}$ for Z, and let $\phi(x + y\sqrt{-1}) = p + q\sqrt{-1}$, where p and q are real, or unit-line, functions of x and y and the unit-line symbols of A, B, &c. When x and y are such that $\phi(x + y\sqrt{-1}) = 0$, let the point of which they are coordinates be called a radical point, single, double, triple, &c., according as there are one, two, three, &c. roots equal to $x + y\sqrt{-1}$. Let the extremity of Z traverse any bounded contour whatsoever in the positive direction of revolution. As it traverses, note the changes of sign in $\frac{p}{q}$, at which the passage is through 0, neglecting the changes at which the fraction passes through ∞. Let k be the number of times that there is a change from + to −, and l the number of times that there is a change from − to +. Then $\frac{1}{2}(k - l)$ is the number of radical points within the contour, on the supposition that each radical point counts as often as the root it indicates occurs.

First, if the theorem be true for each of the contours into which the figure of a larger contour is divided, it is true for the whole. For if all these contours be severally described in the positive direction of revolution, each part, except the external boundary, will be described twice, in two opposite directions. Accordingly, every change from + to − or from − to +, on any part which is described twice, is met by another change from − to + or from + to − when the same line is described in the opposite direction: and this for every thing except the external contour.

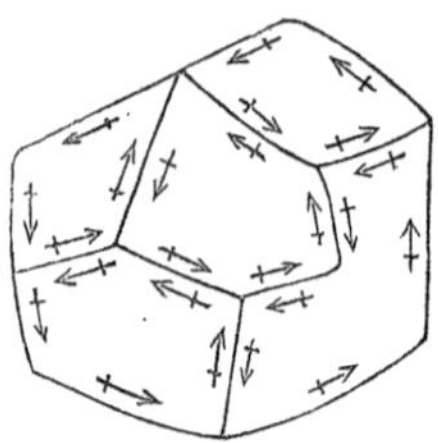

If then we form $k - l$ for each contour, and sum the results, it is the same as if we had formed it for the external contour only.

Suppose the whole contour divided into as many as may be necessary of smaller ones, each of which may be as small as we please. In order that $k - l$ may have any value on a contour, both p and q must vanish on that contour. For if neither vanish, $p \div q$ does not change sign at all, and $k - l$ is but $0 - 0$. If q only vanish, $p \div q$ can only change sign in passing through ∞; and such changes are not to be reckoned as part of k or l. If p only vanish, all the changes of sign are made when $p \div q$ passes through 0, and are all counted: and if *all* be counted, there must be as many from + to − as from − to +; or $k = l$, $k - l = 0$. To give value to $k - l$ there remains only the case in which p and q both vanish. Now if smaller contours be described within those first taken, and smaller within those again, and so on, it must be at last (whatever it may have been at first) that only those contours which have radical points within them, have *both* p and q vanishing on them. For suppose we take one in which there is no radical point, and subdivide it perpetually, and always find internal subdivisions, on the contours of which p and q change sign. We may proceed in this way until the extreme values of x, throughout and within each contour, differ as little as we please, and also the extreme values of y. That is, the values of x and y for which p vanishes approach as nearly as we please to those for which q vanishes, within the contour which has no radical point within it: or $\phi(x + y\sqrt{-1})$ may diminish without limit within that contour; which is absurd.

It is, then, to contours having radical points within them, and no others, that we must look for the possibility of $k-l$ having value. Let the subdivisions be so far increased in number that no one possesses more than one radical point, single or multiple as the case may be. Consider one of them, containing a radical point P, to which s roots belong; and let H be the symbol of OP. Let Q be a point on the contour, and let OQ be Z, and PQ, R. Then $Z = H + R$, and we have

$$\phi Z = \phi(A + R) = MR^s + NR^{s+1} + \ldots$$

because ϕZ is divisible by $(Z - H)^s$, or $\phi(A + R)$ by R^s.

We have then,

$$\phi Z = mr^s \cos(s\rho + \mu) + nr^{s+1} \cos\{(s+1)\rho + \nu\} + \ldots$$
$$+ [mr^s \sin(s\rho + \mu) + nr^{s+1} \sin\{(s+1)\rho + \nu\} + \ldots]\sqrt{-1},$$

$$\text{or} \quad \frac{p}{q} = \frac{m\cos(s\rho + \mu) + nr\cos\{(s+1)\rho + \nu\} + \ldots}{m\sin(s\rho + \mu) + nr\cos\{(s+1)\rho + \nu\} + \ldots}.$$

Let the contour be made so small that the sign of this expression is not affected by r: then it depends at last on that of $\cot(s\rho + \mu)$. In this, while the extremity of Z traverses the contour, ϕ passes through an interval of 2π, and $s\rho + \nu$ through s revolutions. In each of these, the cotangent changes sign from + to −, passing through 0 twice; but the corresponding change from − to + is made in passing through ∞, and must not be regarded. Hence $k = 2s$, $l = 0$; and $\frac{1}{2}(k - l) = s$, the number of roots which the radical point represents.

Since, then, the value of $\frac{1}{2}(k - l)$ for the whole contour is made up of the sum of the values for all the contours of the subdivisions; since no subdivision yields anything except it contain a radical point, when it yields as many units as that point represents roots; the theorem stated follows at once.

Now we have $\phi Z = AZ^n + BZ^{n-1} + \ldots\ldots$,
whence it follows that, if ϕZ be $p + q\sqrt{-1}$,

$$\frac{p}{q} = \frac{az^n \cos(n\zeta + a) + bz^{n-1}\cos\{(n-1)\zeta + \beta\} + \ldots}{az^n \sin(n\zeta + a) + bz^{n-1}\cos\{(n-1)\zeta + \beta\} + \ldots}.$$

Let the contour in question be a circle, with the origin for its centre, of radius Z so great in length as to contain all the radical points; and further, let z then increase without limit.

The sign of $p \div q$ is ultimately always that of $\cot(n\zeta + a)$, which, as before, is shown to yield $2n$ changes from + to −, and no others in which p passes through 0, while ζ passes through 2π. Hence $k = 2n$, $l = 0$, or $\frac{1}{2}(k - l) = n$: that is, every integral expression of the nth degree has neither more nor less than n roots.

Algebraical paradoxes disappear under the application of our significant symbols. The equation $x^{\frac{1}{2}} = c(-x)^{\frac{1}{2}}$ is satisfied independently of x, by $1 = c(-1)^{\frac{1}{2}}$. Change x into $-x$, which it seems we may do, since the equation is now identical, and we have $(-x)^{\frac{1}{2}} = c(x)^{\frac{1}{2}}$. Multiplication gives $x^{\frac{1}{2}}(-x)^{\frac{1}{2}} = c^2 x^{\frac{1}{2}}(-x)^{\frac{1}{2}}$, or $c^2 = 1$. But $-c^2 = 1$.

The explanation is as follows. Let (x, ξ) be the symbol denoted by x in the above equation: if it be real, ξ is a multiple of π; but this matters nothing. Then $-x$ is $(x, \xi + k\pi)$ where k is some odd number. Accordingly, x being a positive arithmetical symbol, we have

$$\left(\sqrt{x}, \frac{\xi}{2}\right) = c\left(\sqrt{x}, \frac{\xi + k\pi}{2}\right),$$

which is satisfied by $1 = c\left(0, \frac{k\pi}{2}\right)$. When we say, change x or (x, ξ) into $-x$, we may, if we like, take a different value of k, and then we have, k' being also an odd number,

$$\left(\sqrt{x}, \frac{\xi + k'\pi}{2}\right) = c\left\{\sqrt{x}, \frac{\xi + (k + k')\pi}{2}\right\},$$

which is satisfied by the same value of c as before. Multiply the equations together, and we have

$$\left(\sqrt{x}, \frac{\xi}{2}\right)\left(\sqrt{x}, \frac{\xi + k'\pi}{2}\right) = c^2\left(\sqrt{x}, \frac{\xi + k\pi}{2}\right)\left\{\sqrt{x}, \frac{\xi + (k + k')\pi}{2}\right\},$$

and k and k' being odd numbers, $k + k'$ is even, say $2k''$. Undoubtedly, then,

one value of $x^{\frac{1}{2}} \times$ one value of $(-x)^{\frac{1}{2}}$

$$= c^2 \{\text{one value of } (-x)^{\frac{1}{2}}\} (\text{one value of } x^{\frac{1}{2}});$$

but we are not now sure of any common factor by which to divide. And the double division is impossible. Let us make it possible with respect to $(-x)^{\frac{1}{2}}$, which must be done by taking k and k' both of the form $4m+1$, or both of the form $4m + 3$.

We have, then, the same form of $(-x)^{\frac{1}{2}}$ on both sides. But then $k+k'$ is in either case of the form $4m+2$, and therefore the division gives

$$\left(\sqrt{x}, \frac{\xi}{2}\right) = c^2 \left\{\sqrt{x}, \frac{\xi}{2} + (2m+1)\pi\right\},$$

which is always satisfied by $c^2 = -1$.

If the successive changes of sign in x be made by one continuous method, say addition of π to the angle, then, starting with one particular form of $x^{\frac{1}{2}}$, say a, we pass successively through a, $a\sqrt{-1}$, $-a$, $-a\sqrt{-1}$, a, &c. If by addition of $-\pi$, then we proceed through a, $-a\sqrt{-1}$, $-a$, $a\sqrt{-1}$, a, &c. And similarly for other roots. The remembrance of the meanings of the symbols will save rules: whatever angle we add to x, we add the mth part of that angle to its mth root.

The same difficulty occurs in treating the equations

$$\phi(-x) = c\phi x, \quad \phi\left(\frac{1}{x}\right) = c\phi x, \text{ \&c.,}$$

all of which may be made to appear to require $c^2 = 1$, as above. But the first is satisfied by

$$\phi x = x^m, \text{ if } (-1)^m = c;$$

and the second by

$$x = (\log x)^m, \text{ if } (-1)^m = c.$$

And the explanation is of the preceding kind in both cases.

As long as a and b are unit-line symbols, and the length of b less than that of a, $\sqrt{(a^2-b^2)}$ is a unit-line symbol, as follows. Let OA and OB be a and b (the reader may supply the diagram), and take $AC = OB$ on the limit line. From O draw a tangent to the circle having centre A and radius AC; let the point of contact be P, and take OQ on the positive part of the unit line $= OP$. Then OQ is $+\sqrt{(a^2-b^2)}$. But when b is of greater length than a, $\sqrt{(a^2-b^2)}$ is the symbol of a line on the axis of direction: O is now within the circle, and if the circle cut the positive axis of direction in R, OR is $+\sqrt{(a^2-b^2)}$. All this is evident from geometry; and nothing in the higher parts of modern geometry is more remarkable than the constant connexion of the shortest semi-chord passing through a point *within* a circle with the length of the tangent drawn from a point *without* it.

The student may investigate for himself the difference of meaning of the following theorem,

$$\sqrt{(a+b)} = \sqrt{\frac{a+\sqrt{(a^2-b^2)}}{2}} \pm \sqrt{\frac{a-\sqrt{(a^2-b^2)}}{2}},$$

(the second term having the sign of b) in the cases in which $a^2 - b^2$ is positive, and those in which it is negative.

CHAPTER VIII.

ON THE ROOTS OF UNITY.

THE roots of unity are really, when algebra is made complete, of an intermediate character between the *quantitative* symbols A, B, &c., and the *directive* symbols + and –. They may be given absolutely to either class of symbols. Thus $(1)^{\frac{1}{n}}$ is $\left(1, \frac{2k\pi}{n}\right)$, k being any positive or negative integer: and thus we find (p. 132) that $(1)^{\frac{1}{n}}$ is the unit of length inclined at any number of nth parts of a revolution. No question, then, that $(1)^{\frac{1}{n}}$ is a perfect particular case of $A^{\frac{1}{n}}$. But if, considering + and – in their directive character, we had chosen* to designate by $(+)^{\frac{1}{n}}$ the prefixed sign of a change of direction which would restore $+A$ back to that form after n performances of its operation: and by $(-)^{\frac{1}{n}}$ a sign of such change of direction as would change $+A$ into $-A$ after n such performances; we might have established the laws of exponents over $(+)^{\frac{1}{n}}$ and $(-)^{\frac{1}{n}}$, and $(+)^{\frac{1}{n}}A$ and $(-)^{\frac{1}{n}}A$ would have had $+A$ and $-A$ for particular cases. But the $(+)^{\frac{1}{n}}A$ and $(-)^{\frac{1}{n}}A$ of the second view are absolutely identical with $(+1)^{\frac{1}{n}} \times A$ and $(-1)^{\frac{1}{n}} \times A$ of the first.

The present chapter treats these roots of unity in a manner which is by no means uncommon; and which in itself involves

* The only point in which I differ from the view taken by my deceased friend, the late D. F. Gregory, one of the most profound thinkers who has ever attended to the subject, lies in this, that he advocated either the necessity or the unavoidable expediency of the second view; and I look upon the two as equally sound, and the choice as a question of convenience which is settled merely by usage.

a full half-leaning to the purely directive definition. The properties of these roots are established on the definition and nothing else: no knowledge of the algebraical forms is demanded, or established for use. The *properties*, for instance, of the forms of $(1)^{\frac{1}{3}}$, their relations to one another, and to the forms of $(1)^{\frac{1}{n}}$ for values of n other than 3, are quite independent of the fact (unknown, it may perfectly well be) that their common forms are 1 and $\frac{1}{2}(1 \pm \sqrt{-3})$. Accordingly, the whole of this chapter might be translated into an algebra of directive signs, of which + and − are mere instances. Thus, that a is a directive sign which repeated n times has the same meaning as +, might be expressed by $a^n = +$; = signifying identity of directive meaning. I first consider the roots of $+1$.

LEMMA. If m and n be integers prime to one another, integers can be found, a and b, such that $mb - na = \pm 1$. Turn $m \div n$ into a continued fraction, and let the approximation preceding the final restoration be $a \div b$. Then, by the property of the successive approximations, $mb - na$ is either $+1$ or -1.

1. Every m^{th} root is an $(mn)^{\text{th}}$ root, m and n being any integers whatsoever. For if $a^m = 1$, then $(a^m)^n = 1$, or $a^{mn} = 1$. The first root $1^{\frac{1}{1}}$ or 1, is a root of every order.

2. Every power of an mth root is an mth root: for if $a^m = 1$, $(a^n)^m = (a^m)^n = 1$, so that a^n is also an mth root. This holds whether n be positive or negative.

3. If m and n be prime to one another, no m^{th} root (except 1) is an nth root. Find a and b so that $mb - na = \pm 1$; if then $a^m = 1$, $a^n = 1$, we must have $(a^m)^b \div (a^n)^a = 1$ or $a^{\pm 1} = 1$, or $a = 1$.

4. If an mth root be an nth root, it is also a kth root, where k is the greatest common measure of m and n. Let $m = km'$, $n = kn'$, m' and n' being prime to one another. Let $m'b - n'a = \pm 1$, then $mb - na = \pm k$. Let α be both m^{th} and n^{th} root: then $\alpha^{mb-na} = 1$, as before, and $\alpha^{\pm k} = 1$, or $\alpha^k = 1$. For instance, we see from (1) that the 8th roots are both 32nd and 40th roots: we now see that the 8th roots are the *only* ones which are both 32nd and 40th roots.

5. There cannot be more than n n^{th} roots. For $x^n - 1$ is $(x - \alpha)(x - \beta)\ldots\alpha, \beta$ being all the roots. As soon as n separate

roots are discovered, the product becomes of the nth degree, and is then identical with $x^n - 1$. And this product cannot vanish except for $x = a$, or β, &c. Nor can any of these roots be equal, since nx^{n-1} has no root except 0.

6. If n be a prime number, and a be one root (not 1), then $1, a, a^2, \ldots a^{n-1}$ are n different roots. For if $a^k = a^l$, k and l being less than l, we have $a^{k-l} = 1$; but n is greater than, and therefore prime to, $k - l$ (being a prime number, and prime to all numbers except its multiples), and an nth root cannot be a $(k - l)^{th}$ root.

7. If m, n, p, &c. be each prime to all the rest, and if a and a' be two different m^{th} roots, β and β' two n^{th} roots, &c., it is impossible that $a\beta\gamma \ldots = a'\beta'\gamma' \ldots$ For an instance, take three classes of roots. Then because a and a' are m^{th} roots, and $a\beta\gamma = a'\beta'\gamma'$, it follows that $\beta^m\gamma^m = \beta'^m\gamma'^m$, and because β^m and β'^m are n^{th} roots, $\gamma^{mn} = \gamma'^{mn}$, therefore $\gamma \div \gamma'$ is an $(mn)^{th}$ root. But $\gamma^p = \gamma'^p = 1$, therefore $\gamma \div \gamma'$ is a pth root, or p and mn being prime to each other, $\gamma \div \gamma'$ is both a pth and an $(mn)^{th}$ root, which cannot be.

8. If n be not a prime number, let P, Q, R, &c. be its prime factors, and let $n = P^p Q^q R^r \ldots$ Then if a be any $P^{p\,th}$ root, β any $Q^{q\,th}$ root, γ any $R^{r\,th}$ root, &c., $a\beta\gamma \ldots$ is an nth root. And all the nth roots can be thus found, and no more. First, $(a\beta\gamma \ldots)^n$ or $a^n\beta^n\gamma^n \ldots$ is $1 \times 1 \times 1 \ldots$, since n is a multiple of P^p, and therefore $a^n = 1$, &c. Therefore $a\beta\gamma \ldots$ *is* an nth root. Next, (by 7), no two such products can give the same n^{th} root, since P^p, Q^q, &c. are prime to each other. Thirdly, since there are P^p $P^{p\,th}$ roots, Q^q $Q^{q\,th}$ roots, &c. the number of combinations of one out of each set is $P^p Q^q \ldots$ or n. Therefore all the varieties of such products give n different n^{th} roots, or all the nth roots and no more.

Accordingly the whole question of finding roots has been reduced to that of prime orders and power-of-prime orders. All the 360th roots, for instance, are found whence the $2^{3\,th}$ $3^{2\,th}$ and 5th roots are found.

9. Every order has some roots which belong to no lower order. If n be a prime number, this is the case with $n - 1$ of

the roots (all except 1). If n be of the form P^p, P being prime, any n^{th} root of a lower order than n must be (7) of the $P^{p-1\,\text{th}}$ order: for, P being prime, P^k is the only form of common measure of P^p and lower numbers, k not exceeding $p-1$. Hence there are $P^p - P^{p-1}$ or $P^{p-1}(P-1)$ of the $P^{p\,\text{th}}$ roots which are of no lower kind. Next, if n be $P^p Q^q R^r \ldots$ and if we take $\alpha\beta\gamma\ldots$ where α is one of the $P^{p\,\text{th}}$ roots which are of no lower kind, &c. then $\alpha\beta\gamma\ldots$ is an nth root of no lower order. For $(\alpha\beta\gamma\ldots)^m = 1$ must give $\alpha^m = 1$, $\beta^m = 1$, &c.: if α^m be not $= 1$, $\alpha^m = (\beta\gamma\ldots)^{-m}$, and α^m being a $P^{p\,\text{th}}$ root, so is $(\beta\gamma\ldots)^{-m}$. But this last is a $(Q^q R^r\ldots)^{\text{th}}$ root, and P^p and $Q^q R^r\ldots$ are prime to each other. Therefore $\alpha^m = 1$, &c. Now since $\alpha^m = 1$, and α is a $P^{p\,\text{th}}$, and no lower root, m has P^p among its factors; since $\beta^m = 1$, &c., m has Q^q among its factors; and so on. Hence m cannot be less than n, or $P^p Q^q R^r \ldots$; while it is obvious that it may be n.

Hence the number of nth roots which are of no lower order is $P^{p-1}Q^{q-1}\ldots \times (P-1)(Q-1)\ldots$: that is, (Arithmetic, p. 196) for every number less than n and prime to it, (1 included) there is an nth root which is no lower root: and all the other n^{th} roots are lower roots.

Let those n^{th} roots which are no lower roots, be called *principal** n^{th} roots. Then there are 4 principal 12th roots: for less than 12, and prime to it, we have 1, 2, 7, 11.

Grant *one* principal root, and all above follows immediately. For if $1, \alpha, \alpha^2, \ldots\ldots \alpha^{n-1}$ be all different, and if we select α^k, in which k is prime to n, then $1, \alpha^k, \alpha^{2k}, \ldots\ldots \alpha^{(n-1)k}$ are all different, and embrace the whole of the first series in a different order. For the succession $1, k, 2k, \ldots (n-1)k$ with each term divided by n, gives the same remainder in no two cases (Arithmetic, p. 195). But if $lk = l'n + r$, $\alpha^{lk} = \alpha^r$: and therefore, in the second series, we see nothing but the first series with its terms altered in order. Thus, if α be a principal 12th root,

* I would have said *primitive* n^{th} roots, but Gauss has used this last word in connexion with the subject of roots. Moreover, it is not that these roots are primitive n^{th} roots, so much as that 'n^{th}' is the primitive ordinal of these roots.

the principal 12th roots are a, a^5, a^7, a^{11}, and if we form the powers of these, dividing by a^{12} whenever it occurs, we have the 12th roots, all of them, arranged in the following sequences:—

$$\begin{array}{cccccccccccc}
1 & a & a^2 & a^3 & a^4 & a^5 & a^6 & a^7 & a^8 & a^9 & a^{10} & a^{11} \\
1 & a^5 & a^{10} & a^3 & a^8 & a & a^6 & a^{11} & a^4 & a^9 & a^2 & a^7 \\
1 & a^7 & a^2 & a^9 & a^4 & a^{11} & a^6 & a & a^8 & a^3 & a^{10} & a^5 \\
1 & a^{11} & a^{10} & a^9 & a^8 & a^7 & a^6 & a^5 & a^4 & a^3 & a^2 & a
\end{array}$$

10. All the powers of an nth root are cyclical. Thus, if a be a principal root, we have cycle of n; for we have $1, a, \ldots a^{n-1}, a^n (= 1), a^{n+1} (= a), \ldots\ldots$ But if a be not a principal root, the cycle is in number sub-multiple of n. If, for instance, n being 12, a be a sixth root, we have $a^6 = 1$, $a^7 = a$, $a^8 = a^2$, &c. The negative powers are only the same cycle repeated backwards; thus $a^{-1} = a^{n-1}$, $a^{-2} = a^{n-2}$, &c.

The most convenient way of considering the roots is by arranging them in reciprocal couples, or from the beginning and end of the cycle. Thus, a being a principal 12th root, we distribute the 12 roots into 1; a and a^{11}, or a and a^{-1}; a^2 and a^{10}, or a^2 and a^{-2}; a^3 and a^{-3}; a^4 and a^{-4}; a^5 and a^{-5}; and lastly, a^6, not a^6 and a^{-6}, for $a^6 = a^{-6}$, and each must be -1. The student must remember not to couple $+1$ and -1.

The cycles of couples have a reverse order, both in the couples and in their succession. Thus the double cycles of 12th roots run thus: 1, a and a^{-1}, a^2 and a^{-2}, a^3 and a^{-3}, a^4 and a^{-4}, a^5 and a^{-5}, a^6 or -1, a^7 and a^{-7}, which is a^{-5} and a^5, a^8 and a^{-8}, which is a^{-4} and a^4, &c.

[Hitherto we have had nothing to distinguish one principal root from another. But when we consider the values of the roots (page 45) we see one pair of roots, both principal, and principal among principals. They are the ones which have the smallest angles in the first revolution, positive and negative: namely,

$$\cos\frac{2\pi}{n} + \sin\frac{2\pi}{n}\cdot\sqrt{-1} \quad\text{and}\quad \cos\left(-\frac{2\pi}{n}\right) + \sin\left(-\frac{2\pi}{n}\right)\cdot\sqrt{-1},$$

$$\text{contained in}\quad \cos\frac{2\pi}{n} \pm \frac{\sin 2\pi}{n}\cdot\sqrt{-1}.$$

These are principal roots, for no power of $\left(1, \frac{2\pi}{n}\right)$ short of the nth is an equivalent of $(1, 0)$. But they are distinguished from all other principal roots, in that they, by their powers, furnish the simplest forms of all the other roots, namely, with angles in the first half-revolutions, positive and negative. They ought to be called *radical* n^{th} *roots*.]

11. We show a mode of forming all the 12th roots whenever we show a mode of proceeding from number to number, in such manner that by casting out 12 whenever it arises, we get the results 0, 1, 2, ... 11, in any order whatever. Thus, beginning with any number, and proceeding by additions of 1, or 5, or 7, or 11, we obtain all the succession 0, 1, ... 11, as in (9). Can we now do this by successive *multiplications?* Trial will give reason to announce, in any case we may take, that, leaving* out 0 (and consequently a^0 or 1), we can always find a multiplier or multipliers which will succeed with a prime number. With 13, for instance, the following multipliers will succeed: 2, 6, 7, 11. Take any number to begin with, say 4; choose 6 as a multiplier; throw out 13 as it arises, and we shall have the succession 4, 11, 1, 6, 10, 8, 9, 2, 12, 7, 3, 5, (cycle complete) 4, 11, 1, &c. Beginning with 1, as most convenient, we have for the 13 13_{th} roots of unity, a, a^6, a^{10}, a^8, a^9, a^2, a^{12}, a^7, a^3, a^5, a^4, a^{11}; which, with 1, complete the list. Of this cycle it is immediately seen that if for a we write any other, as a^5, the cycle is only made to begin in another place, and its successions are uninterrupted. Thus a^5, $(a^5)^6$, $(a^5)^{10}$, $(a^5)^8$, &c., are a^5, a^4, a^{11}, a, &c. That is, we have a method of arranging the roots in recurring cycles such that the substitution of one root for another only disturbs the commencement of the cycle, and not the order in which the roots occur. I return to this subject again.

12. Every function of the n n^{th} roots, or of any of them, which admits of being expanded in integer powers, positive or negative, of them all, is always reducible to the form

$$A_0 + A_1 a + A_2 a^2 + \dots + A_{n-1} a^{n-1},$$

a being a principal root. For when the expansion is made, so

---

* That we must leave out 0 is obvious enough, after what we have seen of it as a starting symbol of addition, as opposed to 1, the starting symbol of multiplication.

that every term is of the form $a^p \beta^q \gamma^r \ldots$, a, β, γ, ... being n^{th} roots, substitution of the values of β, γ, ... in terms of a will give a series of powers of a, which is reduced to the preceding form, since $a^n = 1$, $a^{n+1} = a$, &c. Observe, I here speak of the *form* only: that form may not be fit for calculation, for A_0, A_1, &c., or some of them, may be divergent series.

13. The sum of the nth roots, the sum of the products of every two, of every three, &c., is 0: but the product of all is -1 or $+1$, according as n is even or odd. This follows from the structure of $x^n - 1$, and the theory of equations.

14. The sum of the mth powers of the nth roots of unity is always 0, except where m is n or a multiple of it, positive or negative, and then it is n. For

$$1^m + a^m + a^{2m} + \ldots + a^{(n-1)m} = \frac{1 - a^{nm}}{1 - a^m}.$$

If m be n, or a positive or negative multiple of it, the first side is obviously $1 + 1 + \ldots$ or n. In every other case, a^m is not $= 1$ and a^{nm} is $= 1$: whence the sum of the terms is 0. Better thus, if it were not that proofs of unexpected simplicity are suspicious. Multiply the sum by a^m, it undergoes no alteration except transferring 1^m from the beginning to the end. If the sum be S, we have then $S = a^m S$, or $S = 0$, unless $a^m = 1$.

15. Any symmetrical function of the n n^{th} roots, otherwise real, is real: for every such symmetrical function is a real function of the sum, the products of every two, &c.

16. If in any function of $\sqrt[m]{A}$, $\sqrt[n]{B}$, $\sqrt[p]{C}$, &c., we multiply $\sqrt[m]{A}$ separately by every m^{th} root, $\sqrt[n]{B}$ by every n^{th} root, &c., and introduce every combination of these values into the function, giving $mnp \ldots$ functions in all, and multiply the resulting functions together, the product will be a rational function of A, B, C, &c. For example, $\sqrt{x} + \sqrt[3]{y}$: 1 and -1 are the square roots of unity, let 1, α, α^2, be the cube roots. Then I say that

$$(\sqrt{x}+\sqrt[3]{y})(-\sqrt{x}+\sqrt[3]{y})(\sqrt{x}+\alpha\sqrt[3]{y})(-\sqrt{x}+\alpha\sqrt[3]{y})(\sqrt{x}+\alpha^2\sqrt[3]{y})(-\sqrt{x}+\alpha^2\sqrt[3]{y})$$

is a rational function of x and y.

A rational function of A is known by its presenting the same value, if for $\sqrt[m]{A}$ be substituted in it $\alpha\sqrt[n]{A}$, α being any n^{th} root of unity.

If in the product preceding, which is a symmetrical function of $\sqrt[n]{A}$, $\alpha\sqrt[n]{A}$, ... $\alpha^{n-1}\sqrt[n]{A}$, α being a principal n^{th} root, we substitute $\alpha^k\sqrt[n]{A}$ for $\sqrt[n]{A}$, we have the same function of $\alpha^k\sqrt[n]{A}$, $\alpha^{k+1}\sqrt[n]{A}$, &c....$\alpha^{n+k-1}\sqrt[n]{A}$, or of the same quantities interchanged in order, which, as the function is symmetrical, makes no difference. Hence the product is a real function of A; and so of the rest.

The product of the six factors in the example is $y^2 - x^3$.

17. If α, β, γ, ... be all or some of the nth roots, and if $\phi(\alpha, \beta, \gamma, ...)$ be a function of α, β, γ, ... capable of expansion into $A_0 + A_1\alpha + ... + A_{n-1}\alpha^{n-1}$; then

$$\phi(\alpha, \beta, \gamma, ...) + \phi(\alpha^2, \beta^2, \gamma^2, ...) + ... + \phi(\alpha^{n-1}, \beta^{n-1}, \gamma^{n-1}, ...) = nA_0.$$

Returning to the mode of arrangement in (9), we see that if m and m' be two numbers which, divided by p, leave remainders r and r', say $qp+r$ and $q'p+r'$, the remainder of the product mm' is that of rm'. If then we take a, a^2, ... until we get above p, and then reject the multiple of p, take only the remainder r, go on with ra, ra^2, ... until we get above p again, and proceed as before, we really form the remainders of the successive powers of a. Thus, if we want to know the remainders of the powers of 2 divided by 11, we have but to form the series 2, 4, 8, (16, reject 11) 5, 10, (20, reject 11) 9, (18, reject 11) 7, (14, reject 11) 3, 6, (12, reject 11) 1, 2, 4, 8, &c. Now it is proved, in works on the theory of numbers, that if any prime number be taken, n, there are numbers less than n for which the powers, successively divided by n, yield all the $n-1$ possible remainders before any recur. That one of these should always exist, is enough for our purpose: but, in truth, so many numbers (1 included) as are less than and prime to $n-1$, so many numbers less than n are there for which the powers yield all remainders before any recur. Thus, calling such numbers *primitive** *subordinates* of n, and examining 19, a prime number, we find that 18 has 6 numbers below it, and prime to it. There are then 6 primitive subordinates of 19, and they

* Gauss calls them *primitive roots* of the integer n: but this term would cause confusion, unless the analogies by which it is justified were introduced.

are 2, 3, 10, 13, 14, 15. That is, taking 14, 14^1, $14^2 \dots 14^{18}$, all yield different remainders when divided by 19; so that all the numbers 1, 2, 3, ... 18, are among those remainders. If then a be a principal 19th root of unity, all the 19th roots, except 1, are contained in the set

$$a^{14},\ a^{14^2},\ a^{14^3},\ \dots\ a^{14^{18}},$$

and no one twice. The advantage of this is, that if for a we write another, we only change the commencement of the cycle. Thus, if for a we write a^{14^3}, we only remove the first and second above to the end. This is not the case when we write one for another in the more natural cycle $a, a^2, \dots a^{n-1}$. Remember particularly that the root 1 never enters this series.

Let all the $(n-1)$th roots be known, and let ω be one of them.

Let us consider the expression

$$P = a^m + a^{m^2}\omega + a^{m^3}\omega^2 + \dots\dots + a^{m^{n-1}}\omega^{n-2},$$

m being a primitive subordinate of n. Remember that $\omega^{n-1} = 1$. We see that change of a into a^m is here equivalent to multiplication by ω; change of a into a^{m2} to multiplication by ω^2; and so on. So that P^{n-1} is not affected by writing any other root for a. Hence if P^{n-1} be really constructed by multiplication, it will be found independent of a, or a function of ω only; say Ω. Hence $P = \sqrt[n-1]{\Omega}$ can be expressed. Let the form of it employed be called $\phi\omega$. Do this for each root 1, ω, ω^2, ...ω^{n-2}, ω being a principal $(n-1)$th root: and let $a_1, a_2, a_3, \dots$ be the successive nth roots $a^m, \dots$ We have then, taking the obvious equation when 1 is used for ω,

$$\begin{aligned}
-1 &= a_1 + a_2 + a_3 + \dots\dots\dots\dots\dots + a_{n-1},\\
\phi\omega &= a_1 + a_2\omega + a_3\omega^2 + \dots\dots\dots\dots\dots + a_{n-1}\omega^{n-2},\\
\phi\omega^2 &= a_1 + a_2\omega^2 + a_3\omega^4 + \dots\dots\dots\dots\dots + a_{n-1}\omega^{2(n-2)},\\
&\dots\dots\dots\dots\dots\dots\dots\dots\dots\dots\dots\dots\dots\dots\\
\phi\omega^{n-2} &= a_1 + a_2\omega^{n-2} + a_3\,\omega^{2(n-2)} + \dots\dots\dots\dots\dots + a_{n-1}\omega^{(n-2)(n-2)}.
\end{aligned}$$

From which, by the property in (14), we find

$$\begin{aligned}
(n-1)\,a_1 &= -1 + \phi\omega + \phi\omega^2 + \dots\dots\dots + \phi\omega^{n-2},\\
(n-1)\,a_2 &= -1 + \omega^{n-2}\phi\omega + \omega^{2(n-2)}\phi\omega^2 + \dots\dots\dots + \omega^{(n-2)(n-2)}\phi\omega^{n-2},\\
(n-1)\,a_3 &= -1 + \omega^{n-3}\phi\omega + \omega^{2(n-3)}\phi\omega^2 + \dots\dots\dots + \omega^{(n-2)(n-3)}\phi\omega^{n-2},\\
&\dots\dots\dots\dots\dots\dots\dots\dots\dots\dots\dots\dots\dots\dots\\
(n-1)\,a_{n-1} &= -1 + \omega\phi\omega + \omega^2\phi\omega^2 + \dots\dots\dots + \omega^{(n-2)}\phi\omega^{n-2}.
\end{aligned}$$

Whence it appears that when n is a prime number, the nth roots can be expressed in terms of the $(n-1)^{th}$ roots, and are therefore algebraically determinable when the latter are so.

Writers on this subject give methods of reducing the labour of the preceding: but as my object* is to show the *possibility* only of finding the nth roots when n is a prime number, and the $(n-1)^{th}$ roots are known, I shall content myself with giving at length the determination of the *fifth* roots; 5 being a prime number and the 4th roots known. One primitive subordinate of 5 is 2; and the succession is 2, 4, 3, 1. Hence, a being a fifth root other than 1, and ω a fourth root, the fourth power of $a^2 + a^4\omega + a^3\omega^2 + a\omega^3$ is independent of a. Now, remembering that $\omega^4 = 1$, $\omega^5 = \omega$, &c., the square of the preceding is

$$(a^4 + a^2 + 2) + 2(a^4 + a)\,\omega + (a^3 + a^2 + 2)\,\omega^2 + 2(a^3 + a^2)\,\omega^3,$$

and the square of this will be found, remembering that

$$1 + a + a^2 + a^3 + a^4 = 0,$$

to be $-1 + 4\omega + 14\omega^2 - 16\omega^3$. Let $\omega_1, \omega_2, \omega_3, \omega_4$, be the fourth roots, and Ω_1, &c. the values of the preceding. Then we have

$$\sqrt[4]{\Omega_1} = a^2 + a^4\omega_1 + a^3\omega_1^{\,2} + a\omega_1^{\,3},$$

and similarly for the rest. If $\omega_1 = \sqrt{-1}$, $\omega_2 = -1$, $\omega_3 = -\sqrt{-1}$, $\omega_4 = 1$, we have $\Omega_1 = -15 + 20\sqrt{-1}$, $\Omega_2 = 25$, $\Omega_3 = -15 - 20\sqrt{-1}$, $\Omega_4 = 1$.

We now proceed to discover which of the fourth roots is to be used; nothing being known except that we are to take the same form in all cases. With no restriction, there are $4 \times 4 \times 4 \times 4$, or 256 different systems of equations. One form is determined by the question: $\sqrt[4]{\Omega_4}$ must be -1; for $a^2 + a^4 + a^3 + a = -1$ Hence the form of $\sqrt[4]{1}$ required is that of a principal fourth root moved through an odd number of right angles. Now in the case of $a + b\sqrt{-1}$, each form of the fourth root has all the properties of a principal form; for no one of these fourth roots is a square root. And to $a + b\sqrt{-1}$ and $a - b\sqrt{-1}$ corresponding forms are such as $p + q\sqrt{-1}$ and $p - q\sqrt{-1}$, sym-

---

* The hint of this limitation of object is taken from the late Mr. Murphy's work on the Theory of Equations: but I have not thought it necessary to enter, even so far as Mr. Murphy has done, into the methods of reduction.

metrically disposed with respect to the axis of length. Take such a pair at pleasure, and move them in the same direction through an odd number of right angles, and we have a pair such as $-p+q\sqrt{-1}$ and $p+q\sqrt{-1}$, which are symmetrically disposed with respect to the axis of direction: and such is the pair which must be chosen.

Now if we extract the fourth roots of $-15\pm 20\sqrt{-1}$ by the formula

$$\sqrt{(a\pm b\sqrt{-1})}=\pm\left\{\sqrt{\frac{\sqrt{(a^2+b^2)}+a}{2}}\pm\sqrt{\frac{\sqrt{(a^2+b^2)}-a}{2}}\,.\sqrt{-1}\right\},$$

we shall find them all contained in

$$\pm p\pm q\sqrt{-1},\ \text{and}\ (\pm p.\pm q\sqrt{-1})\,.\,\sqrt{-1},$$

using like signs in the two terms for $-15+20\sqrt{-1}$, and unlike signs for $-15-20\sqrt{-1}$. And $p=\sqrt{\tfrac{1}{2}}(5+\sqrt{5})$, $q=\sqrt{\tfrac{1}{2}}(5-\sqrt{5})$. Choosing a pair symmetrical with respect to the axis of direction, we form the following equations:

$$-p+q\sqrt{-1}=a^2+a^4\sqrt{-1}-a^3-a\sqrt{-1},$$
$$-\sqrt{5}=a^2-a^4+a^3-a,$$
$$p+q\sqrt{-1}=a^2-a^4\sqrt{-1}-a^3+a\sqrt{-1},$$
$$-1=a^2+a^4+a^3+a.$$

Sum these as they stand, and then sum them after multiplication by $-\sqrt{-1}$, -1, $\sqrt{-1}$, 1; -1, 1, -1, 1; $\sqrt{-1}$, -1, $-\sqrt{-1}$, 1. We thus obtain

$$a^2=-\tfrac{1}{4}(\sqrt{5}+1)+\tfrac{1}{2}q\sqrt{-1},\qquad a^4=\tfrac{1}{4}(\sqrt{5}-1)+\tfrac{1}{2}p\sqrt{-1}.$$
$$a^3=-\tfrac{1}{4}(\sqrt{5}+1)-\tfrac{1}{2}q\sqrt{-1},\qquad a=\tfrac{1}{4}(\sqrt{5}-1)-\tfrac{1}{2}p\sqrt{-1},$$

which are well-known values of the fifth roots. Changes of sign in p, or q, or both, have no other effect except different apportionment of the above expressions among the roots a, a^2, a^3, a^4.

The extraction of the square root of $a+b\sqrt{-1}$ is an operation to which Euclid's geometry is competent; it requires only the bisection of an angle, and the determination of a mean proportional, to obtain $\{\sqrt[4]{(a^2+b^2)},\ \tfrac{1}{2}\tan^{-1}b\div a\}$ from $\{\sqrt{(a^2+b^2)},\ \tan^{-1}b\div a\}$.

Hence it follows that wherever n is a prime number, and $n-1$ is a power of 2, the formation of the nth roots of unity is a geometrical* operation, in the ancient sense. Euclid mastered

* This is the discovery of Gauss, and is the most remarkable addition to the power of construction which the ancient geometry has received since the time of Euclid.

the cases $n=3$, $n=5$; the next one is $n=17$, and the next $n=257$.

The theory of the roots of -1 is really contained in that of the roots of $+1$. Since $x^{2n}=1$ is solved both by $x^n=1$ and $x^n=-1$, it follows that all the nth roots of -1 are among the $2n^{\text{th}}$ roots of $+1$. If a be a principal $2n^{\text{th}}$ root of $+1$, we must have $a^n=-1$, $a^{3n}=-1$, &c., and the n n^{th} roots of -1 are seen in $a, a^3, a^5, \ldots a^{2n-1}$. Speaking now of *roots of* -1 *only*, we have the following theorems, answering to some of those in page 148. The student may make a complete list of analogous theorems. Every m^{th} root is an $\{m(2n+1)\}^{\text{th}}$ root. Every odd power of an mth root is an mth root. If m and n be prime to one another, no m^{th} root (except -1, if both be odd) is an nth root: for a $2m^{\text{th}}$ root of $+1$ would then be a $2n^{\text{th}}$ root, which can only happen as to a square root.

If a be a principal $2n^{\text{th}}$ root of 1, it is a principal n^{th} root of -1. For in that case $a, a^3, \ldots a^{2n-1}$ are all different, and only a^n is -1. And there are no other principal n^{th} roots of -1.

Let a be a principal n^{th} root of -1, or $2n^{\text{th}}$ root of $+1$. Then $a, a^3, \ldots a^{2n-1}$ are all different; multiply each by a, and $a^2, a^4, \ldots a^{2n}$ are all different. Nor can any one of the first set be the same as any one of the second: for if $a^{2k}=a^{2l+1}$, we have $a^n=a^{2n(k-l)}$, or $-1=+1$, which is absurd. Therefore there are as many principal n^{th} roots of -1 as $2n^{\text{th}}$ roots of $+1$, and no more.

The sum of the kth powers of the nth roots of -1 is always 0, except where k is a multiple of n. The series $a^k+a^{3k}+\ldots+a^{(2n-1)k}$ is not altered by multiplying by a^{2k}, except by removing the first term to the end: consequently it is 0 except when $a^{2k}=1$, that is, except when k is a multiple of n. If it be an even multiple, the sum is n; if an odd multiple, it is $-n$.

Among the uses which may be made of the roots of unity, the following are remarkable.

An expression may be formed, which goes through recurring periods of changes while x, of which it is a function, takes successive integer values. Let n_x stand for $(\alpha^x+\beta^x+\ldots\ldots)\div n$, α, β, &c., being all the nth roots of $+1$. Then as x changes through $0, 1, 2, \ldots n, n+1$, &c., S_x changes through $1, 0, 0, \ldots 1, 0, \ldots$ Thus $a4_{x-2}$ represents the xth payment of a rent of £a, which is due only every fourth year, the year after next being a year of payment. This is

$$\frac{a}{4}\left\{\cos(x-2)\frac{\pi}{2}+\cos(2x-4)\frac{\pi}{2}+\cos(3x-6)\frac{\pi}{2}+\cos(4x-8)\frac{\pi}{2}\right\},$$

the coefficient of the imaginary part always vanishing in a sum of powers.

This is $\frac{a}{4}\left(1+\cos\pi x-\cos\frac{\pi x}{2}-\cos\frac{3\pi x}{2}\right).$

Again, $a_0 n_x + a_1 n_{x-1} + a_2 n_{x-2} + \ldots + a_{n-1} n_{x-n+1}$ represents an expression which takes the cycle of values $a_0, a_1, \ldots a_{n-1}$ as x passes through $0, 1, \ldots n-1$; and repeats the same while x changes through $n, n+1, \ldots 2n-1$: so that $\phi x = a_r$, when r is the remainder in the division of x by n.

If n_x be formed from the nth roots of -1, the above represents an expression in which the second cycle is formed by changing all the signs, the third by restoring them; and so on.

If ϕx be $A_0 + A_1 x + A_2 x^2 + \ldots$ a finite or infinite series, the roots of unity enable us* to give a finite form to

$$A_m + A_{m+n} x + A_{m+2n} x^2 + \ldots\ldots$$

First, suppose $m < n$: for x write ax, a being one of the nth roots of 1, and multiply by a^{n-m}, forming $a^{n-m}\phi ax$. Do the same for each root and add the results. The theorem on the sums of the powers of the roots then gives

$$\frac{\Sigma a^{n-m}\phi ax}{n} = A_m x^m + A_{m+n} x^{m+n} + A_{m+2n} x^{m+2n} + \ldots$$

Divide by x^m, and write $\sqrt[n]{x}$ for x, and the required result is obtained. If $m > n$, say $m = pn + k\,(k < n)$; find $A_k x^k + A_{k+n} x^{k+n} + \ldots$ and subtract as many of the first terms as are requisite.

If the nth roots of -1 be used, we may in the same manner find $A_m - A_{m+n} x + \ldots$

For instance, let it be required to find

$$1 + \frac{x}{6.7.8.9} + \frac{x^2}{6.7\ldots12.13} + \frac{x^3}{6.7\ldots16.17} + \ldots\ldots$$

---

* In my *Differential Calculus*, I gave this as (to me) new, expressing a doubt that so apparently obvious a method should never have struck any one. I have since found it given by Thomas Simpson, in the Philosophical Transactions, as read Nov. 16, 1758.

Write x^4 for x, and multiply by $x^5 \div 1.2.3.4.5$, and we have selected terms out of ε^x, each fourth one being taken, beginning at $x^5 \div 1.2.3.4.5$. Begin at x, and we should obtain the whole series from

$$\tfrac{1}{4}\{1^3\varepsilon^x + (-1)^3\varepsilon^{-x} + (\sqrt{-1})^3\varepsilon^{x\sqrt{-1}} + (-\sqrt{-1})^3\varepsilon^{-x\sqrt{-1}}\},$$

or
$$\frac{\varepsilon^x - \varepsilon^{-x}}{4} + \frac{1}{2}\sin x = x + \frac{x^5}{1.2.3.4.5} + \ldots\ldots$$

Whence
$$1.2.3.4.5\left\{\frac{\varepsilon^{\sqrt[4]{x}} - \varepsilon^{-\sqrt[4]{x}}}{4\sqrt[4]{x^5}} + \frac{\sin\sqrt[4]{x}}{2\sqrt[4]{x^5}} - \frac{1}{x}\right\}$$

is the value of the required series.

In using this method, it will be best to take a pair of corresponding roots, of the form $\cos\theta \pm \sin\theta\sqrt{-1}$, and work out, in general terms, that part of the sum of the functions which arises from that pair. Suppose for instance, $\phi x = (1 + x)^k$, and that $A_m x^m + A_{m+n} x^{m+n} + \ldots$ is required, m being $< n$. If

$$a = \cos\theta + \sin\theta\sqrt{-1},$$

we have
$$\phi(ax) = (1 + x\cos\theta + x\sin\theta.\sqrt{-1})^k$$
$$= (1 + 2x\cos\theta + x^2)^{\frac{k}{2}}\,\varepsilon^{k\tan^{-1}\frac{x\sin\theta}{1+x\cos\theta}.\sqrt{-1}}.$$

Multiply this by a^{n-m} or by a^{-m}, then change the sign of θ, and add, and we have for the part of $\Sigma a^{n-m}\phi ax$ which depends on the two roots in $\cos\theta \pm \sin\theta.\sqrt{-1}$,

$$(1 + 2x\cos\theta + x^2)^{\frac{k}{2}}.\,2\cos\left\{k\tan^{-1}\frac{x\sin\theta}{1 + x\cos\theta} - m\theta\right\}$$

CHAPTER IX.

SCALAR VIEW OF ALGEBRAICAL SYMBOLS.

It will be admitted that the view of the extended meanings of $A + B$ and $A \times B$ given in pages 118, 119, is a very natural, and even necessary, consequence of the separation of subject matter and operative direction in pages 115, 116: and that it makes the entrance of the extended subject-matter dictate the mode of assigning significance to $A + B$ and $A \times B$. But the transition to A^{B} appears destitute of sufficient obligation to previous suggestion. This chapter, which is above most elementary students, is intended to defend the mode of transition, and to show that the adverse judgment which may be given upon it is partly to be compared to the opinion which a beginner forms upon the law of a series when he has expanded but one or two terms, and which he retracts when he sees those that follow; and partly due to a failure of consistency in algebraic notation.

In $A + B$ and $A \times B$ we see convertible operations, and we take the hint to denote convertible operation by a symbol placed between the subjects: thus $A {}^{\circ} B$, $A * B$, might denote other convertible operations performed with the instruments A, B. Again, we see also a character of ascent, and a connexion: namely, the right of distribution of the higher operation over the *terms*, or separate instruments, of the lower. Let us continue on this hint, and invent an operation which bears to $A \times B$ the same relation as $A \times B$ to $A + B$; and another yet again above the last; and so on. Having symbols for the first two, we keep them: and provide a notation indicative of the degree of ascent, just as we retain the terms *square* and *cube*, and then pass to *third* power, *fourth* power, &c.

Let $A^{\prime\prime\prime}B$, $A^{iv}B$, $A^{v}B$, indicate the successive steps of the ascent, so that our equations of definition are

$A+B=B+A,\ A\times(B+C)=(A\times B)+(A\times C)$	$0+0=0$	$A+0=A$
$A\times B=B\times A,\ A'''(B\times C)=(A'''B)\times(A'''C)$	$1\times1=1$	$A\times1=A$
$A'''B=B'''A,\ A^{iv}(B'''C)=(A^{iv}B)'''(A^{iv}C)$	$\Omega_3'''\Omega_3=\Omega_3$	$A'''\Omega_3=A$
$A^{iv}B=B^{iv}A,\ A^{v}(B^{iv}C)=(A^{v}B)^{iv}(A^{v}C)$	$\Omega_4^{iv}\Omega_4=\Omega_4$	$A^{iv}\Omega_4=A$
&c.	&c.	&c.

Again, in $A+B$ and $A\times B$ we have initial symbols 0 and 1; of which it is the property, as in $0+0=0$, $1\times1=1$, that when no instrument except the initial symbol is used, nothing but the initial symbol results. The equations of the second and third columns are formed by obvious extension.

Now let us denominate by the name of *scalar* function or operation the function which has this property, that its performance on the compound is equivalent to the next lower compound of its performance on the separate terms: so that, if $\lambda_{n-1,n}$ be the symbol of the scalar function connecting $A^{(n-1)}B$ and $A^{(n)}B$, we have

$$\lambda_{n-1,n}(A^{(n)}B) = (\lambda_{n-1,n}A)^{(n-1)}(\lambda_{n-1,n}B).$$

It might at first be supposed that we could have different scalar functions at every transition: but a moment's consideration will show that the perfect accordance of the different symbolic relations would enable us to generalize the scalar system, so as to make all its steps alike, if it should so happen that in any one part of the system we found a scalar function of a more general character than had theretofore appeared. The utmost variety that we can admit is, that in one ascent one particular case should be taken, and in another, another, of the most general form which exists.

Nor is any argument against the above to be derived from the fact of the sequence of operations having a commencement; for there is in truth no commencement. The operation which precedes A^1B or $A+B$ is $A^\circ B$, satisfying

$$A+(B^\circ C)=(A+B)^\circ(A+C).$$

And, understanding any sign + or − in parentheses, in an algebraical sense, we have

$$A_0(B^{(-1)}C)=(A^0B)^{(-1)}(A^0C), \text{ \&c.}$$

The initial symbols may be represented by Ω_0, $\Omega_{(-1)}$, $\Omega_{(-2)}$, &c.

The system of operations is then interminable in both directions. Let λ be the symbol of the scalar function, and let γ represent its inverse function: so that $\gamma\lambda A = \lambda\gamma A = A$, for one value at least. Suppose, in order to assimilate our system to that of algebra, that whatever forms λA may have, γA has but one form; so that $\gamma\lambda A$ is A, though $\lambda\gamma A$ may only have A for *one* of its forms. It would be best, as in algebra, to construct a main system upon one form of λA, and to express the results of other forms of λA in terms of that system. In this trunk-system, we may consider γ and λ as having each only one form; and all the direct operations as having each only one form. It is now clear that if only one of the convertible operations be given, and the scalar function, all are given; and that we have

$$A^{(n+1)}B = \gamma(\overline{\lambda A}^{(n)}\overline{\lambda B}), \quad \Omega_{(n+1)} = \gamma\Omega_n,$$

for $\gamma(\lambda A^{(n)}\Omega_n) = \gamma\lambda A = A$; and this is $A^{(n+1)}\gamma\Omega_n$, whence $\Omega_{n+1} = \gamma\Omega_n$.

The symbols of ordinary algebra, considering the various accidents by which they have attained their positions, have great, but not perfect, consistency. We have $A + B$, $A \times B$, and the scalar function $\log A$; the last derived immediately, by Napier, from the connexion of $A + B$ and AB, and not indirectly* from the exponential function, which he knew nothing of. The exponential function is *out of the system*, strictly speaking: for it will be found that the scale of operations having the indices...... (– II), (– I), 0, I, II, III, IV, &c., is, in the ordinary language, λ being used for log.,

$$\ldots\ldots\lambda^3(\varepsilon^{\varepsilon^{\varepsilon^a}} + \varepsilon^{\varepsilon^{\varepsilon^b}}), \quad \lambda^2(\varepsilon^{\varepsilon^a} + \varepsilon^{\varepsilon^b}), \quad \lambda(\varepsilon^a + \varepsilon^b), \quad a + b,$$

$$\gamma(\lambda a + \lambda b) \text{ or } ab, \quad \gamma^2(\lambda^2 a + \lambda^2 b) \text{ or } \varepsilon^{\lambda a . \lambda b} \text{ or } a^{\lambda b} \text{ or } b^{\lambda a},$$

$$\gamma^3(\lambda^3 a + \lambda^3 b) \text{ or } \varepsilon^{\varepsilon^{\lambda^2 a . \lambda^2 b}}, \text{ \&c.}$$

The initial symbols are $\lambda^3 0$, $\lambda^2 0$, $\lambda 0$, 0, 1, ε, ε^ε, &c.

---

* So that by going back to the sources, we find the logarithm first exhibited as the scalar function by Napier, its inventor, and the trigonometrical system first appearing as *founded on ratios*, in the writings of Rheticus, who first presented it complete.

Accordingly, in A^B, there is not scalar relation between A and B, but between λA and B. And the preceding is not merely justification, but even proof, of the necessity of making $\gamma(\lambda A.\lambda B)$ the definition of the next step to AB, and of making $\gamma(B\lambda A)$ the definition of A^B, our symbolic departure from $\gamma(\lambda A.\lambda B)$.

The origin of A^B is connected with a notion which is developed in the following. Looking at 0 and 1 not as the initial symbols of two distinct operations, but as the initial symbol of an operation (+) and the scalar step of its subject, let successive operations, ϕ and ψ, be defined by the relation that

$$\psi(A+1) = \phi(\psi A, B).$$

That is, let the several definitions be derived from the solutions of

$$\phi(A+1) = \phi A + B, \quad \text{which gives } \phi A = AB,$$
$$\phi(A+1) = B\phi A, \quad \ldots\ldots \quad \phi A = B^A,$$
$$\phi(A+1) = B^{\phi A}. \quad \&c.$$

This last gives a function not capable of finite representation under existing symbols, though we may commence with

$$\phi 1 = a, \quad \phi 2 = B^a, \quad \phi 3 = B^{(B^a)}, \quad \&c.$$

It is neither to be expected nor desired that any substitute should be adopted for A^B; but the more the mind accustoms itself to consider this as a function rather of $\log A$ than of A, the better.

Any two convertible functions of x and y, $\phi(x, y)$ and $\psi(x, y)$ being given, as two consecutive members of a scale, the following condition of distribution must be satisfied,

$$\psi\{x, \phi(y, z)\} = \phi\{\psi(x, y), \psi(x, z)\},$$

and the scalar function must be determined from

$$\lambda\psi(x, y) = \phi(\lambda x, \lambda y),$$

λ being a functional symbol.

Every solution of this system is a chance for the invention of an algebra, in which, $\phi(x, y)$ being denoted by $z + y$ and $\psi(x, y)$ by xy, and $\gamma(\lambda x, \lambda y)$ by $x^{\lambda y}$ or $y^{\lambda x}$, all the laws of ordinary algebra remain good.

In double algebra, the scalar function, in its most general form, is

$$\lambda R = (m + n\sqrt{-1}) \log r + (\mu + \nu\sqrt{-1})\,\theta,$$

the condition $\lambda(r, 0) = \log r$, which is necessary to the complete and unaltered inclusion of single algebra, gives $m = 1$, $n = 0$.

And it will be found on inquiry*, that the adoption of

$$\log r + (\mu + \nu\sqrt{-1})\,\theta$$

is of no effect whatever, except what would in common algebra be called the choice of $\epsilon^{(\mu+\nu\sqrt{-1})}$ instead of $\epsilon^{\sqrt{-1}}$ for a base of angular exponentials.

If a moment's hesitation should arise on the retrograde symbols of the scale, the reader may try the equation

$$A + B\,{}^{\circ}C = (A + B)^{\circ}\,(A + C),$$

$$A + \log(\epsilon^{B} + \epsilon^{C}) = \log(\epsilon^{A+B} + \epsilon^{A+C}).$$

When two successive operations have the required distributive character, that character necessarily attaches to the next one, if formed from the scalar function. Thus, if $A\,'''B$ be defined as $\curlyvee(\lambda A \times \lambda B)$, we have

$$A\,'''\,(B \times C) = \curlyvee\{\lambda A \times (\lambda B + \lambda C)\} = \curlyvee(\lambda A \times \lambda B + \lambda A \times \lambda C).$$

But $P \times Q = \curlyvee(\lambda P + \lambda Q)$ or $\curlyvee(P + Q) = \curlyvee P \times \curlyvee Q$; so that

$$A'''(B \times C) = \curlyvee(\lambda A \times \lambda B) \times \curlyvee(\lambda A \times \lambda C) = A'''B \times A'''C,$$

and so on for the rest.

When the inverse scalar function is used, the regressive system has the same properties as the progressive one with the direct scalar function: for

$$\curlyvee(A^{(n-1)}B) = \curlyvee A^{(n)}\curlyvee B.$$

Hitherto I have said nothing of inverse operations. Let $A_n B$ be the inverse operation of $A^n B$; so that $A^n B_n B = A$. And for $A_{,}B$ and $A_{,,}B$ use $A - B$ and $A \div B$. If any one of the inverse signs follow the rule of signs (p. 103), so does the next. That is, if for instance $\Omega_{3\,,,,}(\Omega_{3\,,,,}A)$ give $\Omega_3{}'''A$ or A, we have $\Omega_{4\,,\text{iv}}(\Omega_4{}^{\text{iv}}A) = \Omega_4{}^{\text{iv}}A = A$. For

$$\lambda(A_{,\text{iv}}B)'''\lambda B = \lambda(A_{,\text{iv}}B^{\text{iv}}B) = \lambda A \text{ or } \lambda(A_{,\text{iv}}B) = \lambda A_{\,,,,}\lambda B,$$

$$\lambda\{\Omega_{4\,,\text{iv}}(\Omega_{4\,,\text{iv}}A)\} = \lambda\Omega_{4\,,,,}\lambda(\Omega_{4\,,\text{iv}}A) = \lambda\Omega_{4\,,,,}(\lambda\Omega_{4\,,,,}\lambda A)$$

$$= \Omega_{3\,,,,}(\Omega_{3\,,,,}\lambda A) = \lambda A, \text{ by hypothesis,}$$

---

* The complete investigation will be found in a paper "On the foundation of Algebra, Part III." in vol. viii. of the *Cambridge Philosophical Transactions*.

or $\Omega_{4_{iv}}(\Omega_{4_{iv}}A)$ is A. And conversely, if the rule of signs be true for $\Omega_{4_{iv}}A$ it is true for $\Omega_{3_{,,,}}A$.

An algebra, similar to ours, requires but the following fundamental basis.

Two consecutive operations, $A + B$, $A \times B$, convertible, so that $A + B = B + A$ and $A \times B = B \times A$, and having the second distributive over the first as in $(B + C) \times A = B \times A + C \times A$.

A scalar operation, λA, having the property

$$\lambda (A \times B) = \lambda A + \lambda B.$$

One starting symbol, 0, wholly ineffective in its own operation, so that

$$0 + 0 = 0, \quad 0 + A = A,$$

An inverse operation, seen in $A - B$, so that $(A - B) + B = A$; and giving $0 - (0 - A) = A$.

Strictly speaking, one operation and its inverse, and the scalar function and its inverse, are sufficient for expression: thus $\curlyvee(\lambda A + \lambda B)$ is sufficient to express $A \times B$. And hence the whole system of scalar functions and starting symbols may be deduced. But the invention of *two* operations, *followed* by that of a scalar function, has been the order of discovery.

The formation of a symbolic system on the seven operations of addition, subtraction, multiplication, division, involution, evolution, and formation of a logarithm, is both redundant and unsymmetrical: but the redundancy is rich in means of expression, and the reduction to symmetry is easy to one practised in the language of algebra as it stands. This last will be best seen by assimilating the notation more closely to that of common algebra. Let 0, $0_{,}$, $0_{,,}$, $0_{,,,}$, &c., be thus defined:

$$0_{,} = 1, \quad 0_{,,} = \varepsilon, \quad 0_{,,,} = \varepsilon^{\varepsilon}, \quad 0_{iv} = \varepsilon^{\varepsilon^{\varepsilon}}, \text{ \&c., } \quad \text{or } 0_{n} = \curlyvee^{n}0.$$

Let $\qquad n_{,} = \varepsilon^{n}, \quad n_{,,} = \varepsilon^{\varepsilon^{n}}, \text{ \&c. } \quad \text{or } n_{k} = \curlyvee^{k}n.$

Let the progressive symbols be $+$, $+_{,}$, $+_{,,}$, &c. and $\times$, $\times_{,}$, $\times_{,,}$, &c. thus connected; $\times$ is $+_{,}$, $\times_{,}$ is $+_{,,}$, &c. Then, the convertible and distributive properties remaining, we have all theorems of ordinary algebra holding good, when any one suffix is placed below $+$ $\times$ and all numerical coefficients. Thus

$$(a +_{,,} b) \times_{,,} (a +_{,,} b) = a \times_{,,} b +_{,,} 2_{,,} \times_{,,} a \times_{,,} b +_{,,} b \times_{,,} b$$

means the following in ordinary language. The first side is

$$\epsilon^{\varepsilon^{\lambda^2(\varepsilon^{\lambda a . \lambda b}) . \lambda^2(\varepsilon^{\lambda a . \lambda b})}};$$

the second side is

$$\epsilon^{\lambda(\varepsilon^{\varepsilon^{\lambda^2 a . \lambda^2 a}}) . \lambda(\varepsilon^{\varepsilon^{\lambda^2 \varepsilon^{\varepsilon^2} . \lambda^2 a . \lambda^2 b}}) . \lambda \varepsilon^{\varepsilon^{\lambda^2 b . \lambda^2 b}}},$$

whence the equation may be easily verified. This chapter may serve to show the necessity of connecting successive operations by the scalar or logarithmic operation, and the ease with which it may be done without any permanent disturbance of established notation.

THE END.

Metcalfe and Palmer, Printers, Cambridge.

WALTON AND MABERLY'S

CATALOGUE OF EDUCATIONAL WORKS, AND WORKS IN SCIENCE AND GENERAL LITERATURE.

☞ *The Works thus marked,* are placed on the List of School-Books of the Educational Committee of the Privy Council.*

ENGLISH.

Dr. R. G. Latham. The English Language.
Fourth Edition. 2 vols. 8vo. £1 8s. cloth.

Latham's Elementary English Grammar, for the Use of Schools. Sixth Edition. 12mo. 4s. 6d. cloth.

Latham's Hand-book of the English Language, for the Use of Students of the Universities and higher Classes of Schools. Third Edition. Small 8vo. 7s. 6d. cloth.

Latham's Logic in its Application to Language.
12mo. 6s. cloth.

Latham's Elements of English Grammar, for the Use of Ladies' Schools. Fcap. 8vo. 1s. 6d. cloth.

Latham's History and Etymology of the English Language, for the Use of Classical Schools. Second Edition, revised. Fcap. 8vo. 1s. 6d. cl.

Abbott's New English Spelling Book; designed to Teach Orthography and Orthoëpy, with a Critical Analysis of the Language, and a Classification of its Elements; on a new plan. Third Edition, with Illustrations. 12mo. 6d.

**Abbott's First English Reader.*
Third Edition. 12mo., with Illustrations. 1s. cloth, limp.

** Abbott's Second English Reader.*
Third Edition. 12mo. 1s. 6d. cloth, limp.

*Newman's Collection of Poetry for the Practice of Elocu*tion. Made for the Use of the Ladies' College, Bedford Square. Fcap. 8vo. 2s. 6d.

*Scott's Suggestions on Female Education. Two Intro*ductory Lectures on English Literature and Moral Philosophy, delivered in the Ladies' College, Bedford Square, London. Fcap. 8vo. 1s. 6d.

GREEK.

*Greenwood's Greek Grammar, including Accidence, Irre*gular Verbs, and Principles of Derivation and Composition; adapted to the System of Crude Forms. Small 8vo. 5s. 6d. cloth.

*Kühner's New Greek Delectus; being Sentences for Trans*lation from Greek into English, and English into Greek; arranged in a systematic Progression. Translated and Edited by the late Dr. Alexander Allen. Fourth Edition, revised. 12mo. 4s. cloth.

Gillespie's Greek Testament Roots, in a Selection of Texts, giving the power of reading the whole Greek Testament without difficulty. With Grammatical Notes and a Parsing Lexicon, associating the Greek Primitives with English Derivatives. Crown 8vo. 7s. 6d. cloth.

Robson's Constructive Exercises for Teaching the Elements of the Greek Language, on a system of Analysis and Synthesis, with Greek Reading Lessons and copious Vocabularies. 12mo., pp. 408. 7s. 6d. cloth.

Robson's First Greek Book. Exercises and Reading Lessons with Copious Vocabularies. Being the First Part of the "Constructive Greek Exercises." 12mo. 3s. 6d. cloth.

The London Greek Grammar. Designed to exhibit, in small Compass, the Elements of the Greek Language. Sixth Edition. 12mo. 1s. 6d. cloth.

Linwood's Lexicon to Aeschylus. Containing a Critical Explanation of the more difficult Passages in the Seven Tragedies. Second Edition, revised. 8vo.

Hardy and Adams's Anabasis of Xenophon. Expressly for Schools. With Notes, Index of Names, and a Map. 12mo. 4s. 6d. cloth.

*Greek Authors. Selected for the Use of Schools; con-*taining portions of Lucian's Dialogues, Anacreon, Homer's Iliad, Xenophon's Memorabilia, and Herodotus. 12mo. 1s. 6d. cloth.

Smith's Plato. The Apology of Socrates, the Crito, and part of the PHAEDO; with Notes in English from Stallbaum, Schleiermacher's Introductions, and his Essay on the Worth of Socrates as a Philosopher. Edited by Dr. WM. SMITH, Editor of the Dictionary of Greek and Roman Antiquities, &c. Third Edition. 12mo. 5s. cloth.

*The Four Gospels in Greek, for the use of Schools, Gries-*bach's Text. Fcap. 8vo. 1s. 6d. cloth.

Tayler's Introduction to the Art of Composing Greek Iambics, in Imitation of the Greek Tragedians, designed for the Use of Schools. 12mo. 2s. 6d.

Æschylus. Prometheus. Wellauer's Text. By GEORGE LONG, A.M. Fcap. 8vo. 1s. 6d. sewed.

LATIN.

New Latin Reading Book; consisting of Short Sentences, Easy Narrations, and Descriptions, selected from Caesar's Gallic War; arranged in Systematic Progression. With a Dictionary. Second Edition, revised. 12mo. 2s. 6d. cloth.

The London Latin Grammar; including the Eton Syntax and Prosody in English, accompanied with Notes. Sixteenth Edition. Fcap. 8vo. 1s. 6d. cloth.

Hall's Principal Roots of the Latin Language, simplified by a Display of their Incorporation into the English Tongue. Sixth Edition. 12mo. 1s. 6d. cloth.

*Allen's New Latin Delectus; being Sentences for Transla-*tion from Latin into English, and English into Latin; arranged in a systematic Progression. Third Edition, revised. 12mo. 4s. cloth.

Robson's Constructive Latin Exercises, for teaching the Elements of the Language on a System of Analysis and Synthesis; with Latin Reading Lessons and Copious Vocabularies. Third and Cheaper Edition, thoroughly revised. 12mo. 4s. 6d. cloth.

Robson's First Latin Reading Lessons. With Complete Vocabularies. Intended as an Introduction to Caesar. 12mo. 2s. 6d. cloth.

Smith's Tacitus; Germania, Agricola, and First Book of the ANNALS. With English Notes, original and selected, and Bötticher's remarks on the style of Tacitus. Edited by Dr. WM. SMITH, Editor of the Dictionary of Greek and Roman Antiquities, etc. Third Edition, revised and greatly improved. 12mo. 5s.

*Hodgson's (late Provost of Eton) Mythology for Versifica-*tion; or a brief Sketch of the Fables of the Ancients, prepared to be rendered into Latin Verse, and designed for the Use of Classical Schools. Fifth Edition. 12mo. 3s. cloth. KEY to Ditto, 8vo. 7s.

Hodgson's Select Portions of Sacred History, conveyed in Sense for Latin Verses. Intended chiefly for the Use of Schools. Third Edition. 12mo. 3s. 6d. cloth. KEY to Ditto, royal 8vo. 10s. 6d. cloth.

Hodgson's Sacred Lyrics, or, Extracts from the Prophetical and other Scriptures of the Old Testament; adapted to Latin Versification in the principal Metres of Horace. 12mo. 6s. 6d. cloth. KEY to ditto, 8vo. 12s. cloth.

Caesar's Helvetic War. In Latin and English, Interlinear, with the Original Text at the End. 12mo. 2s. cloth.

Caesar's Bellum Britannicum. The Sentences without Points. 12mo. 2s. cloth.

Cicero—Pro Lege Manilia. 12mo. 1s. sewed.

Table of Reference to Cicero's Letters, in one Chronological Series. 12mo. 6d. sewed.

HEBREW.

Hurwitz's Grammar of the Hebrew Language. Fourth Edition. 8vo. 13s. cloth. Or in Two Parts, sold separately:—ELEMENTS. 4s. 6d. cloth. ETYMOLOGY and SYNTAX. 9s. cloth.

FRENCH.

Merlet's French Grammar. By P. F. Merlet, Professor of French in University College, London. New Edition. 12mo. 5s. 6d. bound. Or sold in Two Parts:—PRONUNCIATION and ACCIDENCE, 3s. 6d.; SYNTAX, 3s. 6d. (KEY, 3s. 6d.)

Merlet's Le Traducteur; Selections, Historical, Dramatic, and MISCELLANEOUS, from the best FRENCH WRITERS, on a plan calculated to render reading and translation peculiarly serviceable in acquiring the French Language; accompanied by Explanatory Notes, a Selection of Idioms, etc. Fourteenth Edition. 12mo. 5s. 6d. bound.

Merlet's Exercises on French Composition. Consisting of Extracts from English Authors to be turned into French; with Notes indicating the Differences in Style between the two Languages. A List of Idioms, with Explanations, Mercantile Terms and Correspondence, Essays, etc. 12mo. 3s. 6d.

Merlet's French Synonymes, explained in Alphabetical Order, with copious Examples (from the "Dictionary of Difficulties"). 12mo. cloth. 2s. 6d.

Merlet's Stories from French Writers; in French and English Interlinear (from Merlet's "Traducteur"). Second Edition. 12mo. 2s. cl.

Lombard De Luc's Classiques Français, à l'Usage de la Jeunesse Protestante; or, Selections from the best French Classical Works, preceded by Sketches of the Lives and Times of the Writers. 12mo. 3s. 6d. cloth.

GERMAN.

Hirsch. The Return of Ulysses. With a short Grammar and Vocabulary. 12mo. 6s. cloth.

ITALIAN.

Smith's First Italian Course; being a Practical and Easy Method of Learning the Elements of the Italian Language. Edited from the German of FILIPPI, after the method of Dr. AHN. 12mo. 3s. 6d. cloth.

INTERLINEAR TRANSLATIONS.

Locke's System of Classical Instruction. Interlinear TRANSLATIONS. 1s. 6d. each.

Latin.

1. Phaedrus's Fables of Æsop.
2. Virgil's Æneid. Book I.
3. Parsing Lessons to Virgil.
4. Caesar's Invasion of Britain.

Greek.

1. Lucian's Dialogues. Selections.
2. The Odes of Anacreon.
3. Homer's Iliad. Book I.
4. Parsing Lessons to Homer.
5. Xenophon's Memorabilia. Book I.
6. Herodotus's Histories. Selections.

French.

Sismondi; the Battles of Cressy and Poictiers.

German.

Stories from German Writers.

Also, to accompany the Latin and Greek Series.

The London Latin Grammar. 12mo. 1s.6d.
The London Greek Grammar. 12mo. 1s.6d.

An Essay explanatory of the System. 6d.

HISTORY, MYTHOLOGY, AND ANTIQUITIES.

*Creasy's (Professor) History of England. With Illustra-*trations. 1 vol. small 8vo. Uniform with Schmitz's "History of Rome," and Smith's "History of Greece." [*Preparing.*

Schmitz's History of Rome, from the Earliest Times to the Death of COMMODUS, A.D. 192. Ninth Edition. One Hundred Engravings. 12mo. 7s. 6d. cloth.

Robson's Questions on Schmitz's History of Rome. 12mo. 2s. cloth.

Smith's History of Greece, from the Earliest Times to the Roman Conquest. With Supplementary Chapters on the History of Literature and Art. New Edition. One Hundred Engravings on Wood. Large 12mo. 7s. 6d. cloth.

Smith's Dictionary of Greek and Roman Antiquities. By various Writers. Second Edition. Illustrated by Several Hundred Engravings on Wood. One thick volume, medium 8vo. £2 2s. cloth.

Smith's Smaller Dictionary of Greek and Roman Antiquities. Abridged from the larger Dictionary. New Edition. Crown 8vo. 7s. 6d. cloth.

Smith's Dictionary of Greek and Roman Biography and Mythology. By various Writers. Medium 8vo. Illustrated by numerous Engravings on Wood. Complete in Three Volumes. 8vo. £5 15s. 6d. cloth.

Smith's New Classical Dictionary of Biography, Mythology, and Geography. Partly based on the "Dictionary of Greek and Roman Biography and Mythology." Third Edition. 750 Illustrations. 8vo. 18s. cloth.

*Smith's Smaller Classical Dictionary of Biography, My-*thology, and Geography. Abridged from the larger Dictionary. Illustrated by 200 Engravings on Wood. New Edition. Crown 8vo. 7s. 6d. cloth.

Smith's Dictionary of Greek and Roman Geography. By various Writers. Illustrated with Woodcuts of Coins, Plans of Cities, etc. Two Volumes 8vo. £4. cloth.

Niebuhr's History of Rome. From the Earliest Times to the First Punic War. Fourth Edition. Translated by Bishop Thirlwall, Archdeacon Hare, Dr. Smith, and Dr. Schmitz. Three Vols. 8vo. £1 16s.

Newman (F. W.) The Odes of Horace. Translated into Unrhymed Metres, with Introduction and Notes. Crown 8vo. 5s. cloth.

*Newman (F. W.) The Iliad of Homer, Faithfully trans-*lated into Unrhymed Metre. 1 vol. crown 8vo. 6s. 6d. cloth.

*Bathurst (Rev. W. H.) The Georgics of Virgil. Trans-*lated. Foolscap 8vo. Cloth, 4s. 6d.

*Akerman's Numismatic Manual; or, Guide to the Collec-*tion and Study of Greek, Roman, and English Coins. Illustrated by Engravings of many hundred types, by means of which even imperfect and obliterated pieces may be easily deciphered. 8vo. 21s. cloth.

Foster's (Professor) Elements of Jurisprudence. Crown 8vo. 5s. cloth.

BIBLICAL ILLUSTRATION.

Gough's New Testament Quotations, Collated with the Scriptures of the Old Testament in the original Hebrew, and the Version of the LXX.; and with the other writings, Apocryphal, Talmudic, and Classical, cited or alleged so to be. With Notes and a complete Index. 8vo.

PURE MATHEMATICS.

* *De Morgan's Elements of Arithmetic.* Fifteenth Thousand. Royal 12mo. 5s. cloth.

De Morgan's Trigonometry and Double Algebra. Royal 12mo. 7s. 6d. cloth.

De Morgan's Arithmetical Books and Authors. From the Invention of Printing to the Present Time; being Brief Notices of a large Number of Works drawn up from Actual Inspection. Royal 12mo. 2s. 6d. cloth.

* *Ellenberger's Course of Arithmetic, as taught in the Pes*-talozzian School, Worksop. Post 8vo. 5s. cloth.

*** *The Answers to the Questions in this Volume are now ready, price* 1s. 6d.

Mason's First Book of Euclid. Explained to Beginners. Fcap. 8vo. 1s. 9d.

Reiner's Lessons on Form; or, An Introduction to Geo-metry, as given in a Pestalozzian School, Cheam, Surrey, 12mo., with numerous Diagrams. 3s. 6d. cloth.

* *Reiner's Lessons on Number, as given in a Pestalozzian* School at Cheam, Surrey. The Master's Manual. New Edition. 12mo. cloth, 5s. The Scholar's Praxis. 12mo. 2s. bound.

Newman's (F. W.) Difficulties of Elementary Geometry, especially those which concern the Straight-line, the Plane, and the Theory of Parallels. 8vo. cloth, 5s.

* *Tables of Logarithms Common and Trigonometrical to* Five Places. *Under the Superintendence of the Society for the Diffusion of Useful Knowledge.* Fcap. 8vo. 1s. 6d.

Four Figure Logarithms and Anti-Logarithms. On a Card. Price 1s.

Barlow's Tables of Squares, Cubes, Square Roots, Cube Roots, and Reciprocals of all Integer Numbers, up to 10,000. Stereotype Edition, examined and corrected. *Under the Superintendence of the Society for the Diffusion of Useful Knowledge.* Royal 12mo. 8s. cloth.

Wedgwood's Geometry of the First Three Books of Euclid, by direct proof from Definitions alone. With an Introduction on the Principles of the Science. 12mo. 2s. 6d.

MIXED MATHEMATICS.

* *Potter's Elementary Treatise on Mechanics, for the Use* of the Junior University Students. By Richard Potter, M.A., Professor of Natural Philosophy in University College, London. Third Edition. 8vo., with numerous Diagrams. 8s. 6d. cloth.

Potter's Elementary Treatise on Optics. Part I. Con-taining all the requisite Propositions carried to First Approximations, with the construction of Optical Instruments, for the Use of Junior University Students. Second Edition. 8vo. 9s. 6d. cloth.

Potter's Elementary Treatise on Optics. Part II. Containing the Higher Propositions, with their application to the more perfect forms of Instruments. 8vo. 12s. 6d.

Potter's Physical Optics; or, the Nature aud Properties of Light. A Descriptive and Experimental Treatise. 100 Illustrations. 8vo. 6s. 6d.

* *Newth's Elements of Mechanics, including Hydrostatics,* with numerous Examples. By SAMUEL NEWTH, M.A., Fellow of University College, London. Second Edition. Large 12mo. 7s. 6d. cloth.

**Newth's First Book of Natural Philosophy; or an Intro-*duction to the Study of Statics, Dynamics, Hydrostatics, and Optics, with numerous Examples. 12mo. 3s. 6d. cloth.

Kimber's Mathematical Course for the University of London. Second Issue, carefully revised, with a New Appendix. 8vo. 10s.

NATURAL PHILOSOPHY, ASTRONOMY, Etc.

Lardner's Museum of Science and Art. Complete in 12 Single Volumes, 18s., ornamental boards; or 6 Double Ones, £1 1s., cl. lettered.

*** *Also, handsomely half-bound morocco, 6 volumes,* £1 11s. 6d.

CONTENTS.

The Planets; are they inhabited Worlds?
Weather Prognostics.
Popular Fallacies in Questions of Physical Science.
Latitudes and Longitudes.
Lunar Influences.
Meteoric Stones and Shooting Stars.
Railway Accidents.
Light.
Common Things.—Air.
Locomotion in the United States.
Cometary Influences.
Common Things.—Water.
The Potter's Art.
Common Things.—Fire.
Locomotion and Transport, their Influence and Progress.
The Moon.
Common Things.—The Earth.
The Electric Telegraph.
Terrestrial Heat.
The Sun.
Earthquakes and Volcanoes.
Barometer, Safety Lamp, and Whitworth's Micrometric Apparatus.
Steam.
The Steam Engine.
The Eye.
The Atmosphere.
Time.
Common Things.—Pumps.
Common Things.—Spectacles—The Kaleidoscope.
Clocks and Watches.
Microscopic Drawing and Engraving.
Locomotive.
Thermometer.
New Planets.—Leverrier and Adams's Planet.
Magnitude and Minuteness.
Common Things.—The Almanack.
Optical Images.
How to Observe the Heavens.
Common Things.—The Looking Glass.
Stellar Universe.
The Tides.
Colour.
Common Things.—Man.
Magnifying Glasses.
Instinct and Intelligence.
The Solar Microscope.—The Camera Lucida.
The Magic Lantern.—The Camera Obscura.
The Microscope.
The White Ants.—Their Manners and Habits.
The Surface of the Earth, or First Notions of Geography.
Science and Poetry.
The Bee.
Steam Navigation.
Electro-Motive Power.
Thunder, Lightning, and the Aurora Borealis.
The Printing Press.
The Crust of the Earth.
Comets.
The Stereoscope.
The Pre-Adamite Earth.
Eclipses.
Sound.

Lardner's Animal Physics, or the Body and its Functions, Familiarly Explained. 520 Illustrations. 1 vol., small 8vo. 12s. 6d., cloth (see page 15).

Lardner's Animal Physiology for Schools (chiefly taken from the "Animal Physics"). 170 Illustrations. 12mo. 3s. 6d. cloth (see page 15).

* *Lardner's Hand-Book of Mechanics.*
357 Illustrations. 1 vol., small 8vo., 5s.

* *Lardner's Hand-Book of Hydrostatics, Pneumatics, and* Heat. 292 Illustrations. 1 vol., small 8vo., 5s.

* *Lardner's Hand-Book of Optics.*
290 Illustrations. 1 vol., small 8vo., 5s.

* *Lardner's Hand-Book of Electricity, Magnetism, and* Acoustics. 395 Illustrations. 1 vol., small 8vo. 5s.

* *Lardner's Hand-Book of Astronomy and Meteorology,* forming a companion work to the "Hand-Book of Natural Philosophy." 37 Plates, and upwards of 200 Illustrations on Wood. 2 vols., each 5s., cloth lettered.

* *Lardner's Natural Philosophy for Schools.*
328 Illustrations. 1 vol., large 12mo., 3s. 6d. cloth.

* *Pictorial Illustrations of Science and Art. With Ex-*planatory Notes. A Collection of large Printed Sheets, each appropriated to a particular Subject, and containing from 50 to 100 Engraved Figures. Parts I to III, 1s. 6d. each. The size of the sheet is 22 by 28 inches. Any sheet may be purchased separately, price 6d.

Part I. 1s. 6d.	Part II. 1s. 6d.	Part III. 1s. 6d.
1. Mechanic Powers.	4. Elements of Machinery.	7. Hydrostatics.
2. Machinery.	5. Motion and Force.	8. Hydraulics.
3. Watch and Clock Work.	6. Steam Engine.	9. Pneumatics.

* *Lardner's Popular Geology. (From "The Museum of* Science and Art.") 201 Illustrations. 2s. 6d.

* *Lardner's Common Things Explained. Containing:* Air—Earth—Fire—Water—Time—The Almanack—Clocks and Watches—Spectacles—Colour—Kaleidoscope—Pumps—Man—The Eye—The Printing Press—The Potter's Art—Locomotion and Transport—The Surface of the Earth, or First Notions of Geography. (From "The Museum of Science and Art.") With 233 Illustrations. Complete, 5s., cloth lettered.

*** *Sold also in Two Series,* 2s. 6d. *each.*

* *Lardner's Popular Physics. Containing: Magnitude amd* Minuteness—Atmosphere—Thunder and Lightning—Terrestrial Heat—Meteoric Stones—Popular Fallacies—Weather Prognostics—Thermometer—Barometer—Safety Lamp—Whitworth's Micrometric Apparatus—Electro-Motive Power—Sound—Magic Lantern—Camera Obscura—Camera Lucida—Looking Glass—Stereoscope—Science and Poetry. (From "The Museum of Science and Art.") With 85 Illustrations. 2s. 6d. cloth lettered.

* *Lardner's Popular Astronomy. Containing: How to* Observe the Heavens—Latitudes and Longitudes—The Earth—The Sun—The Moon—The Planets: are they Inhabited?—The New Planets—Leverrier and Adams's Planet—The Tides—Lunar Influences—and the Stellar Universe—Light—Comets—Cometary Influences—Eclipses—Terrestrial Rotation—Lunar Rotation—Astronomical Instruments. (From "The Museum of Science and Art.") 182 Illustrations. Complete, 4s. 6d. cloth lettered.

*** *Sold also in Two Series,* 2s. 6d. *and* 2s. *each.*

* *Lardner on the Microscope. (From "The Museum of* Science and Art.") 1 vol. 147 Engravings. 2s.

* *Lardner on the Bee and White Ants. Their Manners* and Habits; with Illustrations of Animal Instinct and Intelligence. (From "The Museum of Science and Art.") 1 vol. 135 Illustrations. 2s., cloth lettered.

* *Lardner on Steam and its Uses; including the Steam* Engine and Locomotive, and Steam Navigation. (From "The Museum of Science and Art.") 1 vol., with 89 Illustrations. 2s.

* *Lardner on the Electric Telegraph, Popularised. With* 100 Illustrations. (From "The Museum of Science and Art.") 12mo., 250 pages. 2s., cloth lettered.

The following Works from "Lardner's Museum," may also be had arranged as described below, handsomely half bound morocco, cloth sides.

Common Things. Two series in one vol.	7s. 6d.
Popular Astronomy. Two series in one vol.	7s. 0d.
Electric Telegraph, and Steam and its Uses. In one vol. .	7s. 0d.
Microscope and Popular Physics. In one vol.	7s. 0d.
Popular Geology, and Bee and White Ants. In one vol. .	7s. 6d.

* *Buff's Familiar Letters on the Physics of the Earth.* Treating of the chief Movements of the Land, the Waters and the Air, and the forces that give rise to them. Edited by Dr. A. W. Hoffman, Professor in the Royal College of Chemistry, London. Fcap. 8vo. 5s., cloth.

* *Twelve Planispheres. Forming a Guide to the Stars for* every Night in the Year. With an Introduction. 9vo. 6s, 6d., cloth.

Bishop's Ecliptical Charts, Hours 0, 1, 2, 3, 4, 5, 6, 7, 8, 9, 10, 11, 13, 14, 19, 20, 21, 22, taken at the Observatory, Regent's Park. 2s. 6d. each.

*Bishop's Astronomical Observations, taken at the Ob*servatory, Regent's Park, during the years 1839—1851. 4to. 12s.

Bishop's Synoptical Table of the Elements of the Minor Planets, between Mars and Jupiter, as known at the beginning of 1855, with the particulars relating to their discovery, etc.; arranged at the Observatory, Regent's Park. On a sheet, 1s.

* *Minasi's Mechanical Diagrams. For the Use of Lec*turers and Schools. 15 Sheets of Diagrams, coloured, 15s., illustrating the following subjects: 1 and 2. Composition of Forces.—3. Equilibrium.—4 and 5. Levers.—6. Steelyard, Brady Balance, and Danish Balance.—7. Wheel and Axle.—8. Inclined Plane.—9, 10, 11. Pulleys.—12. Hunter's Screw.—13 and 14. Toothed Wheels.—15. Combination of the Mechanical Powers.

LOGIC.

De Morgan's Formal Logic; or, The Calculus of Inference, Necessary and Probable. 8vo. 6s. 6d.

Boole's Investigation of the Laws of Thought, on which are founded the Mathematical Theories of Logic and Probabilities. 8vo. 14s.

* *Neil's Art of Reasoning: a Popular Exposition of the* Principles of Logic, Inductive and Deductive; with an Introductory Outline of the History of Logic, and an Appendix on recent Logical Developments, with Notes, Crown 8vo. 4s. 6d., cloth.

ENGLISH COMPOSITION.

Neil's Elements of Rhetoric; a Manual of the Laws of Taste, including the Theory and Practice of Composition. Crown 8vo. 4s. 6d., cl.

MAPS.

Teaching Maps. England, Wales, and Part of Scotland. I. Rivers and Mountains, 6d. II. Towns, 6d.

*Outline Maps. Mercator—Europe—British Isles. Pro-*jection only; Three Maps, folio; single Maps, 4d. each. Projection, with Outline of Country; Three Maps, folio; single Maps, 4d. each.

DRAWING.

*Lineal Drawing Copies for the earliest Instruction. Com-*prising upwards of 200 subjects on 24 sheets, mounted on 12 pieces of thick pasteboard, in a Portfolio. By the Author of "Drawing for Young Children." 5s. 6d.

Easy Drawing Copies for Elementary Instruction. Simple Outlines without Perspective. 67 subjects, in a Portfolio. By the Author of "Drawing for Young Children." 6s. 6d.

Sold also in Two Sets.

SET I. Twenty-six Subjects mounted on thick pasteboard, in a Portfolio. 3s. 6d.

SET II. Forty-one Subjects mounted on thick pasteboard, in a Portfolio. 3s. 6d.

The copies are sufficiently large and bold to be drawn from by forty or fifty children at the same time.

Drawing Materials.

A Quarto Copybook of 24 leaves, common paper, 6d.
Ditto . . Ditto . paper of superior quality, 1s. 3d.
Ditto . . . . 60 leaves, tinted paper, 1s. 6d.
Pencils, with very thick lead, B.B.B., 2s. per half dozen.
Ditto . . Ditto . F. 1s. 6d. ditto.
Drawing Chalk, 6d. per dozen sticks, in a Box.
Port-Crayons for holding the Chalk, 4d. each.

Moore's Perspective: Its Principles and Practice. By G. B. MOORE, Teacher of Drawing in University College, London. In Two Parts, Text and Plates. 8vo. cloth, 8s. 6d.

SINGING.

The Singing Master. Containing First Lessons in Singing, and the Notation of Music; Rudiments of the Science of Harmony; The First Class Tune Book; The Second Class Tune Book; and the Hymn Tune Book. Sixth Edition. 8vo. 6s., cloth lettered.

Sold also in Five Parts, any of which may be had separately.

* I.—*First Lessons in Singing and the Notation of Music.* Containing Nineteen Lessons in the Notation and Art of Reading Music, as adapted for the Instruction of Children, and especially for Class Teaching, with Sixteen Vocal Exercises, arranged as simple two-part harmonies. 8vo. 1s., sewed.

* II.—*Rudiments of the Science of Harmony or Thorough* Bass. Containing a general view of the principles of Musical Composition, the Nature of Chords and Discords, mode of applying them, and an Explanation of Musical Terms connected with this branch of Science. 8vo. 1s., sewed.

III.—*The First Class Tune Book. A Selection of Thirty* Single and Pleasing Airs, arranged with suitable words for young children. 8vo. 1s., sewed.

IV.—*The Second Class Tune Book. A Selection of Vocal* Music adapted for youth of different ages, and arranged (with suitable words) as two or three-part harmonies. 8vo, 1s. 6d.

V.—*The Hymn Tune Book. A Selection of Seventy* popular Hymn and Psalm Tunes, arranged with a view of facilitating the progress of Children learning to sing in parts. 8vo. 1s. 6d.

*** The Vocal Exercises, Moral Songs, and Hymns, with the Music, may also be had, printed on Cards, price Twopence each Card, or Twenty-five for Three Shillings.

CHEMISTRY.

Gregory's Hand-Book of Chemistry. For the use of Students. By William Gregory, M.D., Professor of Chemistry in the University of Edinburgh. Fourth Edition, revised and enlarged. Illustrated by Engravings on Wood. Complete in One Volume. Large 12mo. 18s. cloth.

*** *The Work may also be had in two Volumes, as under.*

* Inorganic Chemistry. Fourth Edition, revised and enlarged. 6s. 6d. cloth.

Organic Chemistry. Fourth Edition, very carefully revised, and greatly enlarged. 12s., cloth.

(Sold separately.)

Liebig's Principles of Agricultural Chemistry ; with Special Reference to the late Researches made in England. Small 8vo. 3s. 6d., cloth.

Wöhler's Hand-Book of Inorganic Analysis ; One Hundred and Twenty-two Examples, illustrating the most important processes for determining the Elementary composition of Mineral substances. Edited by Dr. A. W. Hofmann, Professor in the Royal College of Chemistry. Large 12mo.

Liebig's Hand-Book of Organic Analysis ; containing a detailed Account of the various Methods used in determining the Elementary Composition of Organic Substances. Illustrated by 85 Woodcuts. 12mo. 5s. cl.

Bunsen's Gasometry; comprising the Leading Physical and Chemical Properties of Gases, together with the Methods of Gas Analysis. Fifty-nine Illustrations. 8vo. 8s. 6d., cloth.

Parnell's Elements of Chemical Analysis. Qualitative and Quantitative. Second Edition, 8vo. 9s., cloth.

Parnell on Dyeing and Calico Printing. (Reprinted from Parnell's "Applied Chemistry in Manufactures, Arts, and Domestic Economy, 1844.") With Illustrations. 8vo. 7s., cloth.

Liebig's Chemistry in its Applications to Agriculture and Physiology. Fourth Edition, revised. 8vo. 6s. 6d., cloth.

Liebig's Animal Chemistry; or, Chemistry in its Application to Physiology and Pathology. Third Edition. Part I. (the first half of the work). 8vo. 6s. 6d., cloth.

* *Liebig's Familiar Letters on Chemistry, in its Relations to Physiology, Dietetics, Agriculture, Commerce, and Political Economy.* (*New Edition preparing.*)

Liebig's Researches into the Motion of the Juices in the Animal Body. 8vo. 5s.

A Small Bust of Professor Liebig, in Artificial Ivory. Height 10 inches. Price 15s., or, packed in a box, 16s.

ANIMAL MAGNETISM.

Reichenbach's Researches on Magnetism, Electricity, Heat, Light, Crystallization, and Chemical Attraction, in their relations to the Vital Force. Translated and Edited by Dr. GREGORY, of the University of Edinburgh. In 1 vol. 8vo. 6s. 6d., cloth.

STEAM ENGINE AND RAILWAYS.

* *Lardner on the Steam Engine, Steam Navigation, Roads, and Railways.* Explained and Illustrated. Eighth Edition. With numerous Illustrations. 1 vol. large 12mo. 8s. 6d.

GENERAL LITERATURE.

De Morgan's Book of Almanacs. With an Index of Reference by which the Almanac may be found for every Year, whether in Old Style or New, from any Epoch, Ancient or Modern, up to A.D. 2000. With means of finding the Day of New or Full Moon, from B.C. 2000 to A.D. 2000. 5s., cloth lettered.

Guesses at Truth. By Two Brothers. Cheaper Edition. With an Index. 2 vols. fcap. 8vo. 10s., cloth lettered.

Lyndall's Business as it is, and as it might be. Crown 8vo. 1s. sewed, 1s. 6d. cloth.

THIS Essay obtained the Prize of Fifty Guineas offered by the "Young Men's Christian Association," for the best Essay on the Evils of the present System of Business, with suggestions for their removal.

Herschell's "Far above Rubies." A Memoir of Helen S. Herschell. By her Daughter. Edited by RIDLEY H. HERSCHELL. 12mo. 6s. 6d. cloth.

Rudall's Memoir of the Rev. James Crabb; late of Southampton. With Portrait. Large 12mo., 6s., cloth.

Herschell (R. H). The Jews; a brief Sketch of their Present State and Future Expectations. Fcap. 8vo. 1s. 6d., cloth.

Knox's Christian Philosophy. An Attempt to Display the Evidence and Excellence of Revealed Religion, by its Internal Testimony. Fcap. 8vo. 2s. 6d., cloth.

Leatham's Discovery. A Poem. Fcap. 8vo. 2s. 6d., cloth.

Scott's Love in the Moon. A Poem. With Remarks on that Luminary. Fcap. 4to. 5s. 6d., cloth gilt.

Common-Place Books.

The LITERARY DIARY, or Complete Common-place Book, on the Plan recommended by Locke, with an Explanation, and an Alphabet of Two Letters on a Leaf. Post 4to., ruled throughout and half-bound, 8s. 6d.

A POCKET COMMON-PLACE BOOK. With Locke's Index. Post 8vo., half-bound. 6s. 6d.

Frere's Embossed Books for the Blind.

OLD TESTAMENT.

Genesis, 8s.
Exodus, 7s.
Deuteronomy, 7s.
Joshua, 4s. 6d.
Judges, 4s. 6d.
1 Samuel, 6s.
2 Samuel, 5s. 6d.
1 Kings, 6s.
2 Kings, 6s.
Ezra and Nehemiah, 5s. 6d.
Job, 5s.
Psalms, Part I., 6s. 6d.
Psalms, Part II., 6s.
Proverbs, 5s. 6d.
Isaiah, 7s. 6d.
Daniel, Esther, and Ruth, 5s. 6d.

NEW TESTAMENT.

Matthew, 6s.
Mark, 5s. 6d.
Luke, 7s.
John, 5s. 6d.
Acts, 7s.
Romans to Corinthians, 6s.
Galatians to Philemon, 5s. 6d.
Hebrews to Revelation, 7s.

Psalms, Part I. 6s. 6d. } Prayer Book
Psalms, Part II. 5s. 6d. } Version.
Church Catechism, 1s.
Olney Hymns, 2s.
Morning Prayers, 2s.
Two Shoemakers, 6s.
Shepherd of Salisbury Plain, 2s.
Grammar for the Blind, 2s.
Five Addresses to those who wish to go to Heaven. 1s. 6d.

Frere's Works on Prophecy.

GENERAL STRUCTURE OF THE APOCALYPSE. 8vo. 2s., cloth.

THREE LETTERS ON THE PROPHECIES: viz. on the true place of the Seventh Seal; the Infidel Individual Antichrist; and Antiochus Epiphanes as a supposed subject of Prophecy. 8vo. 2s.

EIGHT LETTERS ON THE PROPHECIES: viz. on the Seventh Vial; the Civil and Ecclesiastical Periods; and on the Type of Jericho. 8vo. 2s.

GREAT CONTINENTAL REVOLUTION; marking the Expiration of the "Time of the Gentiles," A.D. 1847-8. 8vo. 2s. 6d.

Introductory Lectures, Delivered in University College, London.

SESSION 1856-57.

ON SELF-TRAINING BY THE MEDICAL STUDENT. By E. A. PARKES, M.D., Professor of Clinical Medicine in the College. Fcap. 8vo. 1s.

SESSION 1828-29.

Dr. Conolly on the Nature and Treatment of Diseases. 1s.
Professor Galiano on the Spanish Language and Literature. 1s.
Dr. Grant on Comparative Anatomy and Zoology. 1s.
Dr. Mühlenfels on the German and Northern Languages and Literature. 1s.
Dr. Smith on Medical Jurisprudence.

SESSION 1829-30.

Professor Amos on English Law. 1s.
Dr. Malkin on History.

SESSION 1830-31.

Professor Bennett on Anatomy. 1s.
Professor De Morgan on Mathematics, Natural Philosophy, and Chemistry. 1s.
Professor Thomson on Medical Jurisprudence. 1s.
Professor Amos on English Law. 1s.

SESSION 1831-34.

Dr. Grant on Medical Education. 1s.
Professor Malden on the Greek and Latin Languages. 1s.
Dr. Quain on Anatomy. 1s.

SESSION 1837-38.

Professor De Morgan on the Establishment of the University of London. 1s.

SESSION 1838-39.

Professor Kidd on the Nature and Structure of the Chinese Language. 1s. 6d.
Professor Malden on the Introduction of the Natural Sciences into General Education. 1s.
Professor Pepoli on the Language and Literature of Italy. 1s.
Professor Carey on the Study of English Law. 1s. 6d.

SESSION 1840-41.

Professor Creasy on History. 1s.
Professor Latham on the English Language and Literature.

SESSION 1842-43.

Professor Donaldson on Architecture. 1s. 6d.

SESSION 1844-48.

Mr. George on Dental Surgery. 1s.
Professor Newman on the Relations of Free Knowledge to Moral Sentiment. 1s.
Professor Ramsay. Passages in the History of Geology. 1s.
Professor Marsham on Law. 1s.

SESSION 1848-49.

Professor Scott on the Academical Study of a Vernacular Literature. 1s.
Professor Ramsay. (Second Lecture) Passages in the History of Geology. 1s.

SESSION 1849-50.

Professor Williamson — Development of Difference the Basis of Unity. 1s. 6d.

SESSION 1850-51.

Professor Erichsen on Surgery. 1s.
Professor Foster on Natural Law. 1s.

SESSION 1851-55.

Professor Chapman on the Relations of Mineralogy to Chemistry and Physics. 1s.
Professor Masson on College Education and Self-Education.

PHARMACY.

Mohr and Redwood's Practical Pharmacy. The Arrangements, Apparatus, and Manipulations of the Pharmaceutical Shop and Laboratory. Illustrated by 400 Engravings on Wood. 8vo.

ANATOMY.

Dr. Quain's Anatomy. Edited by Dr. Sharpey and Mr. ELLIS, Professors of Anatomy and Physiology in University College, London, Illustrated by 400 Engravings on Wood. New and Cheaper Edition. 3 vols., small 8vo. £1 11s. 6d., cloth lettered.

Ellis's Demonstrations of Anatomy. Being a Guide to the Dissection of the Human Body. By GEORGE VINER ELLIS, Professor of Anatomy in University College, London. Fourth Edition. Small 8vo. 12s. 6d., cloth.

Quain and Wilson's Anatomical Plates. A Series of Anatomical Plates in Lithography. Cheap Issue, at the following very low prices:—

		Plain. £	s.	d.	Coloured, £	s.	d.
MUSCLES. 51 Plates	Cloth	1	5	0	2	4	0
VESSELS. 50 Plates	,,	1	5	0	2	0	0
NERVES. 38 Plates	,,	1	1	0	1	14	0
VISCERA. 32 Plates	,,	0	17	0	1	10	0
BONES AND LIGAMENTS. 30 Plates	,,	0	17	0	1	0	0

The Work complete, containing 201 Plates, 2 vols. royal folio, half-bound morocco, price £5 5s. plain; £8 8s. coloured.

SURGERY.

Erichsen's Science and Art of Surgery. Being a Treatise on Surgical Injuries, Diseases, and Operations. By JOHN ERICHSEN, Professor of Surgery in University College. Second Edition, revised and much enlarged. Illustrated by 400 Engravings on Wood. 1 vol. 8vo. £1 5s.

Quain on Diseases of the Rectum. By Richard Quain, F.R.S., Professor of Clinical Surgery in University College. Second Edition, with additions. 12mo. 7s. 6d., cloth.

Morton's Surgical Anatomy of the Principal Regions of the Human Body. 1 vol. royal 8vo., coloured Plates and Woodcuts. £1 1s., cloth.

*A Portrait of the late R. Liston, Esq., Surgeon to Uni-*versity College Hospital. Drawn on Stone by GAUCI, from the original by EDDIS. Price 3s. 6d.

PHYSIOLOGY.

Kirkes's Handbook of Physiology. Third Edition. One Vol., small 8vo., with Illustrations on Steel and Wood. 12s. 6d., cloth lettered.

Lardner's Animal Physics; or, the Body and its Functions familiarly Explained. Upwards of 500 Illustrations. 1 vol., small 8vo. 12s. 6d.

CONTENTS.—General View of the Animal Organization; Bones and Ligaments; Muscles; Structure, of the Lower Animals; Nervous System; Circulation; Lymphatics; Respiration; Digestion; Assimilation, Secretion, the Skin, Animal Heat; Senses; Touch; Smell; Taste; Vision; Hearing; Voice; Development, Maturity, Decline, Death.

Lardner's Animal Physiology for Schools (chiefly taken from the "Animal Physics"). 170 Illustrations. 12mo. 3s. 6d. cloth.

MEDICINE.

Pharmacopœia ad usum Valetudinarii Collegii Universitatis Londinensis, Accommodata. 18mo. 1s. 6d., cloth.

Walshe on the Nature and Treatment of Cancer. By W. H. WALSHE, M.D., Professor of Medicine in University College, Physician to University College Hospital, and Consulting Physician to the Hospital for Consumption and Diseases of the Chest. 1 vol., 8vo., with Illustrations.

Walshe's Practical Treatise on the Diseases of the Lungs, Heart, and Aorta; including the Principles of Physical Diagnosis. Second Edition. 12mo. 12s. 6d., cloth.

Ballard's Artificial Digestion as a Remedy in Dyspepsia, Apepsia, and their Results. Fcap. 8vo. 1s. 6d.

Ballard on Pain after Food; its Causes and Treatment. 12mo. 4s. 6d., cloth.

Ballard's Physical Diagnosis of Diseases of the Abdomen. 12mo. 7s. 6d.

*Jones on Gravel, Calculus, and Gout. Chiefly an Applica-*tion of Professor Liebig's Physiology to the Prevention and Cure of these Diseases. By H. BENCE JONES, M.D., Cantab., F.R.S., Fellow of the Royal College of Physicians, London; Physician to St. George's Hospital. 8vo. 6s.

Murphy on Chloroform, its Properties and Safety in Childbirth. 12mo. 1s. 6d., cloth.

MATERIA MEDICA.

Garrod's Essentials of Materia Medica, Therapeutics, and the Pharmacopœias. For the Use of Students and Practitioners. By ALFRED BARING GARROD, M.D., Professor of Materia Medica and Therapeutics in University College, London. Fcap. 8vo. 6s. 6d., cloth.

GYMNASTICS.

Chiosso's Gymnastics, an Essential Branch of National Education. By CAPTAIN CHIOSSO, Professor of Gymnastics in University College School. 8vo. 1s. 6d.

*Chiosso's Gymnastic Polymachinon. Instructions for Per-*forming a Systematic Series of Exercises on the Gymnastic and Callisthenic Polymachinon. 8vo. 2s. 6d., cloth.

www.ingramcontent.com/pod-product-compliance
Lightning Source LLC
LaVergne TN
LVHW010603110826
845149LV00003B/751